Citizen SCIENCE

Citizen SCIENCE

How Ordinary People Are Changing the Face of Discovery

CAREN COOPER

The Overlook Press
New York, N. Y.

This edition first published in hardcover in the United States in 2016 by
The Overlook Press, Peter Mayer Publishers, Inc.

NEW YORK
141 Wooster Street
New York, NY 10012
www.overlookpress.com
For bulk and special sales, please contact sales@overlookny.com,
or write us at the above address.

Cataloging-in-Publication Data is available from the Library of Congress

Book design and typeformatting by Bernard Schleifer
Manufactured in the United States of America
FIRST EDITION
ISBN 978-1-4683-0867-9
2 4 6 8 10 9 7 5 3 1

To my daughters, my strongest characters,
my complete alphabet, Abby and Zoe

CONTENTS

"Make a difference about something other than yourselves."

—Toni Morrison

Citizen SCIENCE

INTRODUCTION

Marking the Tide

The effect of this state of things is to make the medical profession a conspiracy to hide its own shortcomings. No doubt the same may be said of all professions. They are all conspiracies against the laity.

—GEORGE BERNARD SHAW

I N 1837, AT THE ANNUAL MEETING OF THE ROYAL SOCIETY OF LONDON, William Whewell finally received the recognition he had been working hard to achieve for years. Whewell was in his early forties, ambitious in his career. He was not yet married, but would eventually have two wives (sequentially, not simultaneously). He had just completed his book, *The History of Inductive Sciences*, which would go on to become highly influential in the shaping of science as a profession so beneficial to humanity that it should receive financial support from governments. At the time, the Royal Society's annual meeting traditionally honored two high-achieving individuals. (Today it honors three each year.) By a twist of fate, in 1837 Whewell stood alone in the spotlight: that year he was the one and only recipient of the Royal Medal (also called the Queen's Medal). Britain's highest scientific accolade has been awarded about four hundred times since its inception in 1826, though fewer than ten women have been recipients. In case the name Royal does not adequately convey the importance and prestige of this award, it might be useful to know that Charles Darwin received it twice. The Royal Medal from the Royal Society is the epitome of entry into a highly exclusive professional club for scholars who churn out new knowledge for the benefit of humanity. Their motto: *Nullius in verba* (Take nobody's word for it).

I emphasize the lofty and exclusive reputation carried by the Royal Medal because its award to Whewell for this particular research accomplishment is, in a certain way, an enormous irony. Whewell received this tribute for his scholarly contributions to the understanding of ocean tides in a project he called the "great tide experiment." Yet he accomplished this research by relying on what amounted to almost a million observations collected by thousands of ordinary people living in coastal towns. Volunteers included dockyard officials, sailors, harbormasters, local tide table makers, coastal surveyors, professional military men, and amateur observers. From small notes collected by thousands while going about their daily lives, Whewell crafted and tested his theories.

Whewell was a pioneer in what today we call citizen science. It was without doubt borne of necessity: Whewell did what was needed to fulfill his research agenda, and relying on volunteers along the edge of the seas improved the quality of his research. Like the conductor of a global orchestra, he coordinated thousands of people in nine nations and colonies on both sides of the Atlantic in the synchronized measurement of tides. In our present era, when millions sit down and watch the Super Bowl at the same time, this synchrony may at first seem trivial. So I challenge you to arrange, without the help of a phone or the Internet, to meet even just five friends at an appointed time at a café next month and see how many of them show up! In Whewell's case, he arranged for volunteers at more than 650 tidal stations to follow his specific instructions for measuring tides around the clock at *exactly* the same points in time for two weeks in June 1835. Synchronous measurements on beaches and harbors worldwide were key because he hoped to draw cotidal lines across the ocean: a connect-the-dots effort from port to port. He wondered whether the timing of low tides in, say, London corresponded with high tides in, say, Boston. He found that the ocean was more complex than soup in a bowl being rocked back and forth.

Today citizen science not only fulfills research goals but also helps what is termed informal science education (that is, learning that takes place without a textbook or classroom). Whewell did not intend to increase the science literacy of the populace, nor did he gain a leg up in the eyes of the Royal Society because his insights were a collective effort. He received the honor because his insights were damned important. Great Britain was an empire, and by dominating ocean travel it could monopolize global trade. Figuring out the complexity of the tides was tricky and essential for moving from port to port. The basic knowledge that the moon influenced the tides had been accepted since the time of Galileo, but that abstract cause and effect was not useful in the daily prediction of the heights of local tides. Local tide charts were prepared by those with homespun secrets passed down from generation to generation; their reliability was great locally, but could not be extrapolated to other ports. By pioneering citizen science, Whewell also created a new field of science he called tidology, and was at the forefront of efforts to bring the study of tides away from celestial studies and down to earth (or the beach) in order to make reliable real-world predictions at any port. Yet, even after Whewell's royal accomplishment, tide charts remained difficult to refine. More than a hundred years after Whewell's work, an unexpected high tide in 1953 on the River Thames resulted in the drowning deaths of three hundred people.

In Whewell's day, the term *citizen science* did not exist. If it had, he would have been the person to coin it, because he was the go-to person for scientific jargon; he is responsible for terms like *ion, anode,* and *cathode* even though he was not involved in their discovery. When Whewell was laying the groundwork for his "great tide experiment," he coined the term *scientist*. First, in 1833, it was a chivalrous move: he created the term to avoid having to refer to Mary Sommerville as a "man of science." This allows us to truthfully claim that, technically, the first to sport the title of scientist was a woman. Then, in 1834, he realized a larger necessity for the term: Whewell, Sommerville, and others employed at universities and pursuing scholarly inquiry were polymaths with interests in astronomy, physics, biology, chemistry, and more. "We need very much a name to describe a cultivator of science in general," Whewell said. "I should incline to call him a Scientist. Thus we might say that as an *artist* is a musician, painter, or poet, a *scientist* is a mathematician, physicist, or naturalist." It took several decades for the term to fall into common use, and many more before male and female scientists were commonplace.

The term *citizen science* was needed many years later for similar reasons. To paraphrase Whewell, we might say that *citizens* are those with rights and responsibilities to participate in some larger collective (such as governance), and *citizen scientists* are thus people exercising their rights and responsibilities to participate in collective scientific endeavors. Participation in the process of governance involves adding one's values, opinions, and perspectives to decision making; participation in the process of science involves adding one's observations and amateur expertise to making new knowledge. In the former, one casts ballots; in the latter, one submits data.

The term *citizen science* illustrates the point that birdwatchers who share checklists of birds are doing the same thing, scientifically speaking, as volunteer riverkeepers who measure water quality and amateur astronomers who keep watch for supernovas. Citizen scientists participate in science through different hobbies or concerns, not necessarily through their professions.

Recognizing the importance of activities unrelated to our profession is unusual in the present era. Starting in childhood, we prepare for professions: we play firefighter, we play detective. We structure our lives and identities around a variety of occupations. Some of our surnames come from professions, like Baker, Brewster, Cooper, Gardner, Hunter, Miller, and Smith.

Science as an occupation is a fairly new concept. In Whewell's time, and for centuries before him, science was less often a career and more often an extravagance for the wealthy. Science didn't necessarily require expensive equipment or training, but it did require an abundance of spare time; most who pursued a scientific endeavor did so as an elite hobby. Charles Darwin was not hired as a scientist on the HMS *Beagle*; he was a companion to Captain Robert FitzRoy and a "gentleman naturalist" traveling the world before moving on to his planned occupation as a parson. Gregor Mendel, who deciphered hereditary traits through a series of experiments breeding pea plants, was a monk. The roots of science have always been in leisure time, spare time, or spiritual time.

The roots have also been imperialistic. Another irony of Whewell's research is that, by today's standards, we might use his engagement of volunteer data collectors as evidence that he was an egalitarian scientist, a man ahead of his time in partnering with the laity. Instead, Whewell was very much of his time, focused on aggregating scientific observations to a single place of intellect, the United Kingdom. When his contemporaries, like Darwin, explored the world and collected specimens, the specimens were only considered useful if brought back to the collection of the British Museum.

As Darwin wrote in his diary, in reference to *Megatherium* fossils, "the only specimens in Europe are the King's collection at Madrid, where for all purpose of science they are nearly as much hidden as if in their primaeval rock." Darwin, Whewell, and their contemporaries never envisioned an egalitarian science; they could scarcely conceive of an international one. Whewell's ideas about the process of science supported imperialist notions of his day.

Whewell, the person who relied on citizen science to achieve his highest honor, helped delineate science as an exclusive profession with specific norms and procedures for valid discovery. After *The History of Inductive Sciences*, he wrote *The Philosophy of Inductive Sciences*, carefully considering questions about knowledge production such as who creates new knowledge and who has access to it. Whewell and his peers built upon the system of their predecessors and strengthened the idea that, through their exclusive network, scientists were the overlords of knowledge production. He viewed scientific observations (such as tide levels) as pearls, and induction as the rational mental processes by which intelligent minds (scientists) could string the pearls together to form a necklace. In truth, however, he was unable to assemble that necklace on his own. Having amassed almost a million observations from volunteers, he hired calculators—that is, men who understood calculus—to crunch the numbers. He referred to his calculators and volunteers as "subordinate labourers."[1]

---

[1] I've dared several project leaders to try "Dear subordinate laborers" in their correspondence with participants, in honor of Whewell, but none have accepted the challenge.

As such, Whewell was the one who got the Royal Medal because he helped divide the world into two kinds of people: those who create knowledge and those who don't.

Whewell and his colleagues eventually became known as scientists. They shared the necklaces of knowledge that they crafted among themselves by publishing papers, speaking at annual meetings, and honoring each other with awards and recognition. They constructed a system of peer validation in which authoritative knowledge was confirmed. They often worked in secrecy, waiting years before sharing their findings, communicating with each other and not with the public except when they needed observations. For example, Darwin drew observations from around the world through written correspondence, which includes over fifteen thousand letters. What of all the others outside the established profession of science who shared observations when asked to? Their contributions were unacknowledged and unnamed, instead attributed to the solo science practitioners.

Today citizen science is bringing back and asserting recognition for collective styles of inquiry and transforming the imperial system of science that emerged with Whewell. The revival of the "it takes a village" approach is helping to dispel the negative connotations that have been associated with scientific interests. When I was young, the common view was that the ability to grasp the power of new knowledge only came to madmen (and -women) with Einsteinian hair or big-brained pencil-necks in white lab coats. *Boffins*, *dorks*, *dweebs*, *eggheads*, *geeks*, and *nerds*: these are words that reinforce the notion that science is not for everyone, that scientists are separate from society. Consequently, scientists are both social outcasts and game changers. They are either smart and dangerous or smart and dull, but they are never ordinary.

I was ordinary, and I imagined scientists actually were, too, in their daily lives. I was not as interested in big laboratories as I was in naturalists. I rationalized that natural scientists just worked in distant lands. *National Geographic* explorers went to remote places to make incredible discoveries: George Schaller went to the Himalayas to study snow leopards; Jane Goodall went to Tanzania to study chimpanzees. From my perspective, these scientists were changing the world by bringing back discoveries from extraordinary places.

After working for more than a decade as a scientist in the field of citizen science, I have come to the conclusion that my early image of a scientist—working alone in faraway places, bringing back discoveries that could better the world—was pure childhood fantasy. But an out-of-date paradigm can be like prescription medicine: it can become less effective, even potentially harmful, after it's expired. Let's toss the time-worn archetype from our cabinet of remedies. The human race faces a host of big problems that scientists alone can't solve: overpopulation, climate change, emerging diseases, deforestation, mountaintop removal, great garbage patches in the ocean, and other urgent, contested issues. Scientists, as ordinary people or extraordinary heroes, cannot cure these woes while their methods are cordoned off, available only to a few. To generate effective solutions, we need to relocate the scientific process of discovery away from its isolation. We need to foster the ability of ordinary people to make use of this powerful force in the mainstream of society.

This book is an exploration of the world of citizen science. A portrait of citizen science is both too intricate and too big to view all at once, and so I will reveal it in pieces, discipline by discipline. Citizen science has long histories in meteorology, my own field of ornithology, entomology, and astronomy. In part 1, I dedicate a chapter to each of these fields to illustrate areas old enough to have benefited from citizen science well before the advent of the Internet and smartphones.

I became curious about what citizen science looked like in other fields and around the globe. In time I saw citizen science in every discipline that I reviewed, including relatively new and rapidly developing fields of study. This is covered in part 2, with biochemistry, microbiology, and conservation biology. I saw citizen science on every continent and in too many fields to cover comprehensively in this book: mammalogy, fisheries, limnology, botany, forestry, archeology, animal behavior, immunology, and neuroscience, to name a few. Today every scientific discipline worth its salt has added citizen science to its toolbox. Yet, as I dug deeper, I saw that it wasn't just scientists that needed citizen science to advance their research agendas; many kinds of circumstances led communities to need access to the authority of science, because achieving their own agendas required attaining new, reliable knowledge. I explore the heart of the matter in part 3, with chapters on marine biology, geography, and public health. These stories illustrate that citizen science enables all of us to collectively manage the health and well-being of ourselves and our planet.

The stories that I relate here about citizen science range from awe-inspiring to heartwarming to gut-wrenching and, on rare occasions, impressively nonextraordinary. They are real-life stories about your friends, neighbors, and relatives: people just like you. Perhaps they may even be about you or the person next to you in line at the grocery checkout counter. I may describe some discoveries that your contribution has made possible, even if you were not aware of the impact of your efforts. How many people who dutifully type the distorted phrase required by web security tools to authenticate that the visitor is a person (rather than a computer program with inferior optical recognition skills) realize they are helping, piecemeal, to transcribe words in a book?[2] In a similar way, the cumulative results of many small acts of curiosity are part of a beautiful scientific revolution.

How do we know the migratory routes of birds? How do we know that today they migrate earlier due to climate change? How do we know that the waste from industrial hog farms causes negative consequences for human health? How do we know that monk seals have attempted to recolonize the eastern Mediterranean Sea? How do we know that there is a bestiary of bacteria living in our belly buttons? There is an enormous wealth of knowledge that we may implicitly as-

---

[2] To prevent automated spammers, website security tools will often present visitors with a distorted word that optical character recognition software cannot read. Visitors gain access to the site by typing these words, which proves they are human, not robots. In the case of the web security tool ReCaptcha, one of the words functions as a security check, and the other word is one that was indecipherable for character recognition software used during massive efforts to digitize libraries of books. Typing in the word helps books become digitized.

sume comes exclusively from people working in the scientific profession. But actually, we all have each other to thank.

The topics of study differ. The locations change. The ways that people and scientists collaborate are unique. What remains the same is that citizen science simultaneously creates two interlocking keys needed to solve our big problems: (1) reliable knowledge of what can be done, and (2) social capital to make it happen. *Social capital* refers to the social networks, cohesion, and individual investment in community that make democracy work better. Each chapter in this book highlights scientific discoveries for which we have citizen science to thank, and each chapter uncovers the social benefits that citizen science brings to participants, communities, and society.

In writing this book I was guided by the desire to understand citizen scientists. Who are they? How do they contribute? Why do they do it? And, do they realize the impact they're having? The stories in this book will help us reimagine who carries out science, how it is carried out, where it happens, and whom it serves. We'll revisit these questions at the end of the book and find that citizen science provides answers quite different from Whewell's: we won't find elite scientists and subordinate laborers. In the following chapters, we'll meet individuals ahead of their time, who model for us the lifestyle of empowered global citizens of a sustainable future, and a new breed of publicly engaged scientists helping make it happen. Instead of pinning hopes on a hierarchical system of science, we can turn to an egalitarian system of citizen science for its potential to pull humanity through some of our biggest problems and solve some of our most enduring mysteries.

PART 1
Hobbies of Discovery

IN THIS SECTION I FEATURE CITIZEN SCIENCE ENDEAVORS AND TRADITIONS that originated many decades ago. While you almost certainly have some familiarity with weather bugs, birdwatchers, butterfly enthusiasts, or stargazers, I hope an in-depth look at their stories will overturn a few common misconceptions. For starters, citizen science is clearly not a new phenomenon, even though the phrase was coined fairly recently. Citizen science projects thrived before the advent of the Internet and smartphones. Yet even though citizen science was possible without these technologies, like most things in life, these and other new technologies make the practice of citizen science easier, faster, and more efficient.

I hope the stories in this section will counter the common misconception that citizen science is free simply because it involves volunteers. Yes, volunteers save money because their labors are free, and those savings add up. But there is no free lunch; though volunteers freely share observations with researchers, researchers are then saddled with the costs of computer infrastructure to turn the observations into useful data and archive them in perpetuity. Furthermore, the best way to show appreciation for volunteer contributions is to develop online systems to help volunteers make meaning and use of the collective data.

Finally, I hope you will take away from the following chapters that, contrary to the beliefs of skeptics and critics, the quality of data from citizen science is useful for many purposes. When citizen science is a well-designed collaboration between scientists and the public, new knowledge is coproduced. This means that citizen science is not simply outreach, extension, or environmental education. With the help of science enthusiasts, researchers can explore boundaries of discovery and find answers that would otherwise not be attainable.

CHAPTER 1

Meteorology
NOAA and the Flood

Alone we can do so little; together we can do so much.
—HELEN KELLER

IN MARCH 2003 A SNOWSTORM WAS FORECAST FOR COLORADO'S FRONT Range. This was business as usual for the Denver area. "Our local meteorologists give forecasts for ski resorts and the Denver city limits. That doesn't tell me anything," explains Vivian Kientz. "If a true up-slope was coming, I'd have it bad." Kientz lives a mere fifteen miles from the outskirts of Denver, on the north-facing slope of a mountain. The term *upslope* refers to an air system traveling along the ground that is forced to rise when it meets a mountain slope; when the air rises, it cools and water vapor condenses into rain. Because precipitation changes based on the terrain, and the terrain varies greatly over short distances, a reliable forecast practically needs to be tailored for every valley and ridge. For such a personalized forecast, Kientz visits the web-site of the National Weather Service, a division of the National Oceanic and Atmospheric Administration (NOAA), and enters her exact latitude and longitude.

"Everyone was clueless," she tells me in 2014, recalling the events of 2003, when a true upslope was forecast that would bring several feet of snow. She warned all her friends and neighbors, who ran to the gro-cery store to buy candles, food, beer, and other provisions.

Meanwhile, Kientz went to the hardware store to supplement her weather-monitoring supplies. Her precipitation gauge is a standard-issue double cylinder. The narrow inner cylinder, with demarcated lines like a ruler, catches rain caught by a wider funnel on its top; it holds up to an inch of rainwater. The outer cylinder is four inches wide and is used to catch snowfall by removing the funnel and inner cylinder at the right time (when merited according to the forecast). For the expected blizzard, Kientz needed a longer outer cylinder and bought a several-foot stretch of four-inch-wide stovepipe. She also needed a snow board; not a snow-board for descending a mountain or doing tricks in a half-pipe, but simply a sixteen-inch-square piece of plywood, painted white, to serve as a sampling area.

In October 2002 Kientz had seen an ad in the local newspaper seeking volunteers for the Community Collaborative Rain, Hail & Snow Network; this mouthful goes by the acronym CoCoRaHS (pronounced "cocoa rahs"). Since joining, she has not missed a single day of collecting precipitation data. A Tennessee native transplanted to Colorado twenty-five years ago, she has grown accustomed to snow at any time of year—even summer. She grew up with rain, "five inches at a time back in western Tennessee," she explains on the phone from her home in Colorado, where she hardly gets rain at all. Most weather systems cross North America from the west heading east; most precipitation is dropped on the west side of the Rocky Mountains. Consequently, the east side of the range, called the rain shadow, is quite dry. As Kientz relates, "When people here say, 'Oh, it rained today,' I'm like, 'What? You call that rain?'" Despite having multiple sclerosis, which has Kientz intermittently in and out of a wheelchair, she always shovels a path through the snow to take CoCoRaHS readings. She declares herself an expert at "knowing the weather for my one spot on earth." She's enough of an expert to know that the growing season is too short for gardens. She has a greenhouse measuring ten feet by thirty feet, where she grows hundreds of cactus, orchids, and exotic plants. She knows exactly when her driveway will freeze, and when it will thaw. She knows when she'll be able to dig herself out, and when to call someone with a plow.

The day before the storm, Nolan Doesken, a Colorado climatologist and the founder and director of CoCoRaHS, had sent an e-mail that explained to participants that their mission, should they choose to accept it, would involve extra work to get good measurements for this particular snowfall. Since snow accumulates at different densities, participants collect measurements on snow depth and water content; this is called the snow water equivalent ratio. For a blizzard, participants need a yardstick (because a one-foot desk ruler is too short) to measure depth on the snow board. Then they collect one column of snow in their four-inch-wide gauge. They turn the gauge upside down, drive it like a cookie-cutter into the snow on the board, invert the board and gauge as though turning a cake from its baking pan onto a cooling rack (or putting a spatula under it), and then wipe the board clean and let more snow accumulate. For a big storm with more than twelve inches of new snowfall expected, they have to repeatedly measure snow as it accumulates (or, if they want to sleep all night, like Kientz they buy a long stovepipe to get a deep sample in the morning). Participants bring each snow sample inside to slowly melt and then pour the liquid into their rain gauge to measure the amount. Typical measurements during the 2003 storm were ratios around eight to one (where eight inches of snow melted to one inch of water), which is wet, heavy, sticky snow. Snowboarders and skiers like fluffy, powdery snow with a ratio of fifteen to one or higher.

The storm arrived in the early afternoon of March 17, 2003, and ended on March 19, the day US troops invaded Iraq. Snow fell for three days in what meteorologist Doug Wesley calls "a climatological anomalous snowstorm," and it came down fast; hundreds of roofs collapsed under the weight of so much wet snow. Highways were closed, and people were stranded at Denver International Airport and surrounding ski resorts. Thousands of drivers sought refuge in Red Cross shelters and hotels.

Kientz bundled up and went outside to take measurements of the snow with the dedication of a school kid memorizing multiplication tables. She is a relatively tall woman, at five feet eleven inches, but the snow eventually surpassed her height. All told, she measured seventy inches (six feet) of snow in her one spot on earth, while reports indicated that more than five feet of heavy snowfall covered most of the region. In the weeks following, Denver residents filed over $100 million in insurance claims.

Most people recognized the storm's silver lining: the massive precipitation brought an end to an extreme drought, at least in that region.

Another silver lining was the research opportunity it presented. It was a perfect storm for a citizen science event because (1) meteorologists knew something big was brewing; (2) they had an army of trained volunteers in the bull's-eye; and (3) they had the ability (via e-mail) to communicate with volunteers to prepare and encourage them to do the extra hard work. "In 2003, people were still reading e-mails," muses Doesken (each year, about 60 percent of CoCoRaHS participants return from the previous year, but volunteer fatigue is a general issue in citizen science), "and people stepped up and took this as a challenge."

The legacy of harnessing the power of dedicated volunteer weather observers in the United States can be traced back to as early as 1776. When Thomas Jefferson wasn't busy penning the Declaration of Independence, he was devising a plan to deputize one person in each county in Virginia with a thermometer, a wind vane, and instructions to log observations of temperature and wind direction twice daily. Jefferson experimented with the most high-tech weather devices of his era, including rain gauges and barometers; he is the United States' original weather bug. He was diligent in record keeping and, like Kientz, abhorred gaps in his data.

Yet Jefferson was not setting a new trend by observing weather. The tradition of collecting weather data is as old as civilization. The oldest known written weather records are inscribed on "oracle bones" from the Shang dynasty in China (eighteenth to twelfth century BCE). Shang diviners used sharp knives to engrave ox bones and turtle shells with weather records. They would first inscribe questions on the bones or shells, apply heat until the bone or shell cracked, and then interpret the crack to make a prediction. The questions were sometimes about weather, and the predictions were early weather forecasts. Sometimes the diviners would follow up and inscribe the actual weather outcome—called the verification—on the same pieces of bone or shell (which are now considered valuable as artifacts). Sadly, the records are not a complete representation of time; they were never intended to be daily records, as such documents commonly would be today. But today these oracle bones, about 150,000 of which remain in collections, are of interest to climate researchers. (Unfortunately, prior to about 1900, when such records were discovered, the oracle bones were mistaken for Pleistocene fossils, called dragon bones, and ground up and taken as medicine: the plastrons were used to treat malaria, and poultices made from ox bones were used to heal knife wounds).

Later dynasties kept records of unusual weather, as well as phenological records of the blooming dates of flowering trees. By around 100 BCE, the Chinese had techniques to measure rainfall and snowfall, but there are no extant descriptions of how this was done, so it will remain an ancient Chinese secret. They used weathercocks for wind direction, and wind flags—poles with feathers—to estimate wind speed. They even measured humidity in a coarse way based on the heaviness of charcoal.

Though weather data preceded formal science, a scientific dispute motivated Jefferson's particular desire for county-by-county precipitation data: he wanted evidence that he hoped would refute a European claim, the theory of degeneracy: the reprehensible idea that the temperature and humidity of the New World produced animals that were smaller, weaker, and just plain inferior to their European counterparts. The impetus to douse this claim was strong, because the theory was developed by a Frenchman, Georges-Louis Leclerc, Comte de Buffon, who—remarkably, for his time—classified humans as part of the animal kingdom. Thus his theory amounted to the French smacking a white glove against the cheek of the new Americans. As a relative newcomer to the New World, Jefferson had little data to support counterclaims of superiority. It was the patriotic chip on the founding father's shoulder that sparked his realization that the scientific strength of a country lay in its people.

With daily weather data Jefferson planned to develop his own theory of climate. Unfortunately the Revolutionary War took priority over a systematic statewide plan for data collection. Nevertheless, from 1776 to 1816, President Jefferson and many of his recruits (including explorers Lewis and Clark) kept a near complete series of weather observations. Ultimately, Jefferson used the weather data, including that of five years he spent in France, to show that America had a higher sunny-to-cloudy-day ratio than Europe.

Despite a century of interest and instrumentation, forecasting the weather remained locked down to local observations and folklore. Rules of thumb included such forecasts as "If rain falls during an east wind, it will continue a full day" and "When deer are in gray coat in October, expect a severe winter." My favorite are folklore rhymes, like "Clear moon, frost soon" and "Hark! I hear the asses bray, we shall have some rain today." Folklore was replaced with real observations by 1845, when use of the telegraph became widespread. The roots of a federal weather forecasting program started simply, as people in Virginia could telegraph people in New York to say what weather was on the way. Joseph Henry, the first secretary of the then new Smithsonian Institution, started organizing the relay of weather communications. "A system of observation which shall extend as far as possible over the North American continent," Henry wrote, with the vision that "the extended lines of the telegraph will furnish a ready means of warning the more northern and eastern observers to be on the watch from the first appearance of an advancing storm." By 1848, volunteer observers were actively recruited and telegraph companies allowed weather reports to be transmitted to the Smithsonian free of charge. By 1850, over 150 volunteers were reporting regularly. By 1860, daily telegraphed weather reports were printed in the *Washington Evening Star*. It was decades of telegraph use to relay messages of bad weather and war tragedies that prompted Western Union to introduce the singing telegram in the 1930s in hopes of employing the medium to bring happy news.

The telegraphed reports in the 1860s were about observed, not predicted, weather. Cooperation among states stalled during the Civil War years, but soon after it was viewed as the government's duty to provide forecasts to prevent weather-related tragedies. An act of Congress in 1870, signed into law by President Ulysses S. Grant, required the secretary of war to take meteorological observations across the Great Lakes, the Gulf of Mexico, and the Atlantic coast. In 1872, another act of Congress extended the service to the entire United States "for the benefit of commerce and agriculture." The Signal Service Corps led the charge and would fly different flags (for example, a red square with a dark center meant a storm, and two of those meant a hurricane) in the middle of towns as their way of letting people know what sort of weather was coming. Subsequent problems with embezzlement and other scandals led President Benjamin Harrison to move the national weather service from the War Department to the Department of Agriculture in 1890.

President Harrison charged the Department of Agriculture's new civilian agency, the Weather Bureau,[3] with relying heavily on volunteer observers. These volunteer efforts were the precursor to what is now better known as the National Weather Service's Cooperative Weather Observer Network,[4] which draws about one million volunteer hours annually at twelve thousand sites across all fifty states. There is approximately one network station for every five hundred square miles, and without the network we would often get caught in the rain and we'd know much less about climate trends. This program is part of why Kientz was able to use her latitude and longitude to get a fine-scale forecast for her side of the mountain in Colorado before the big storm. Cooperative weather observers receive certificates of appreciation every five years; those who observe for sixty years or longer receive a letter signed by the president of the United States. Recipients of a letter from the president include Edward Stoll, who observed for seventy-six years

---

[3] The Weather Bureau became the National Weather Service in 1970.
[4] The Australian government's Bureau of Meteorology has almost identical systems that began in 1908.

in Arapahoe, Nebraska, Ruby Stufft, who observed for seventy years in Elsmere, Nebraska, and Richard Hendrickson, of Bridgehampton, on Long Island, New York, who began as a cooperative weather observer in 1930, when he was eighteen years old, and has continued for over eighty years.

The CoCoRaHS was borne not from the wisdom of a past president or an innovation such as the telegraph but of a tragic error in forecasting. In 1997, meteorologists in Fort Collins, Colorado, misjudged the severity of an imminent rainstorm. Unlike the biblical character Noah, NOAA's examination of the heavens did not reveal a prophecy of a flood—just heavy rains. But a flood came nonetheless. A small area near the foothills at the base of the Rockies received 14.5 inches in just a few hours, while areas close by received no more than 2 inches. Only three weather stations existed in the entire community. The disaster led to the death of five people, millions of dollars in property damage and, eventually, the formation of CoCoRaHS.

Radar instruments, ground-based or on satellites, are great for predicting rainfall over large areas, but precipitation can be highly localized. It can rain, snow, or hail on one side of the street and be dry on the other. One county can experience floods while the neighboring county is in the midst of a drought. It goes against our concept of progress, but when it comes to precipitation, nothing high-tech in the sky can beat a low-tech gauge on the ground. For example, remote sensors cannot confidently distinguish rain from snow. Only data collected by home-based volunteers can provide the coverage and quality needed for the research opportunity presented by the anomalous snowstorm of 2003.

By braving the cold, Kientz and other volunteers provided evidence of remarkably localized patterns in snowfall and water content that helped weather forecasters realize the extreme geographic microvariation in precipitation. Doesken explained that all meteorologists recognize that precipitation is variable (annually and geographically), but admits that he did not fully appreciate the variation until seeing the CoCoRaHS data. This lesson led him to quit his common practice of interpolation of contour maps because he no longer feels that he has sufficient density of data. A thorough understanding of what we see in hindsight helps improve meteorological foresight. Wesley was the lead author of a paper that amounted to a postmortem of the storm and why it snowed where it did. The detailed dissection revealed how to refine the traditional understanding of how terrain shapes storm properties.

Wesley, a meteorologist in Alaska, regales me with sinister-sounding meteorological terms like *cold-air damming* and *diabatic cooling*. In Colorado, prevailing weather comes from the west. The tall Rocky Mountains slow the movement of the prevailing winds and the stalled clouds typically drop their precipitation on the west side. The west side of the Rockies is wet during the winter months; the east side (the area of the rain shadow) is dry. Conditions were different for the 2003 storm, which brought moist air from as far away as New Orleans and the Gulf of Mexico and drove it up the east slope of the Rockies. But, as Wesley likes to emphasize, the easterly flow became blocked. The air should flow along the ground, and pressures should push it up and over the Rockies. But this particular easterly flow cooled so quickly that no amount of pressure was going to force it over the mountains. Instead the easterly flow billowed in front of the mountains and accumulated for six to twelve hours. This billowing mass of air acted like invisible mountain terrain and, in effect, displaced the storm by about twenty miles to the east, over cities and towns in the foothills and plains, rather than upon the upper reaches of the mountains. Precipitation fell as though the steepest terrain were near Interstate 25 in Denver rather than in the mountain terrain of Breckenridge.

When I ask Wesley about data quality, he says, "Sheer numbers outweigh problems. CoCoRaHS is a gold mine of data." He did quality control checks, eliminating extreme high and low outliers: "Would you rather have five perfect observations, or one hundred, of which eighty are good?"

When I speak to Doesken, he laughs. "CoCoRaHS participants are younger than third graders, and over ninety. We can't expect the same quality of data from all. The best tool for quality control is redundancy." That means that if a second grader and a ninety-one-year-old neighbor get the same measurement, then the data are likely trustworthy.

CoCoRaHS participants are kindred spirits in terms of being bitten by the weather bug. Doesken knew that participants would step up their efforts to measure the storm because no one likes data gaps. He says he admires their dedication, and has frequently explained to the uninitiated that many participants will "check their rain gauge every day until they literally can't move." He used to say "until their dying day," until he was told that the phrase was insensitive, but the perceived insensitivity stems from a truth: it is in fact common for people to collect data until they die.[5] A case in point is Ned Somerville, who e-mailed Doesken, "I have only been with you only about a few years. I was a meteorologist in my military days, am 75 years, and am in my final weeks with cancer. From here on out, I will do the best I can to get the data in on time, but sometimes that will not be possible. Thank you. It's been a fun job."

Another came from a participant with an unusual name, Howard P. Howard, who wrote to Doesken a year before he died, "I wanted to thank you for recognizing the effort [of] many of us old/ailing volunteers. I don't know how many others agree with me, but for many of us the days of once being the boss, the superintendent, president, foreman or any job that one has worked to fine tune a job over a life time and suddenly you are faced with retirement and or ill health, it is a very scary part of life. But being able to be affiliated with CoCoRaHS gives one a chance to do something worthy and for that I'm grateful."

A social scientist might require more convincing, but receiving a stream of these e-mails over the years has led Doesken to believe that citizen science participation (and the sense of duty it carries) improves life—and may very well prolong it too.

---

[5] This fact is useful to coroners trying to determine the time of death of a participant. At least twice CoCoRaHS has been contacted by a coroner's office for details about the deceased's last weather record in order to narrow the course of events leading up to the death of observers who lived, and died, alone.

• • •

I was raised to believe that weather makes for trifling, boring small talk and is therefore a safe subject to raise when conversing with strangers at the bus stop. To the contrary, weather is one of the strongest influences on human existence. Rain has shaped the daily life and plans of people since the very first parades, weddings, baseball games, picnics, perms, and suede shoes. I'd bet *Homo sapiens* has always been trying to predict how far it's possible to stray from the mouth of the cave without something like an umbrella. Weather is so integral to our existence that we equate it with our moods: fights in movies take place when there's lightning, tragedies during rain, and romance alongside rainbows.

Wilbur and Orville Wright grew up in Dayton, Ohio, but made the first powered flights in Kitty Hawk, North Carolina, because they needed the town's steady coastal winds (and its soft sand for landings). France and California are known for their vineyards because the moderate weather permits grapes to fully ripen. Weather has repeatedly played a hand in history. Monsoons will deter battles, snowstorms may block supply routes, and droughts will surely bring famine.

Weather can exert its impact on us in subtle ways. Antonio Stradivari relied on more than skill to make his famous, high quality, Stradivarius violins. He selected only wood that had grown slowly and evenly (with low density and a high modulus of elasticity)—specifically, wood from trees that grew in the weather of the Little Ice Age, a period spanning roughly the fifteenth and sixteenth centuries. Patterns of weather dictate where particular plants and crops can grow, and ultimately determine where we live: where our civilizations thrive and where they fail.

I meet another weather bug, David Herring, a modern-day deputy of Jefferson, as I pass through Richmond, Virginia. Grinning, he holds up a short pole with several spires. One spire has tiny cups that rotate horizontally in the wind; another has a compass wind vane; a third has a box that holds a rain gauge that automatically empties, and a fourth, which looks like a shock absorber, is for measuring temperature and humidity. Each spire is made of white plastic, looking like the battler armor of the imperial stormtroopers in the movie *Star Wars*. Herring is showing me his home weather station. "I'm not a gizmo type of person, but this is my pride and joy," he says.

Herring explains that he thinks about weather every day. His home weather station electronically transmits to a computer tablet display that he keeps on his kitchen table. In his den Herring keeps a ship's barometer, a weather gauge from the 1940s, and a German weather house in the design of a miniature alpine chalet. At first I thought the chalet was a cuckoo clock, but instead of having a pendulum timing the emergence of a cuckoo every hour, the weather house operates as a barometer: when the barometric pressure drops, a man with an umbrella rotates out; when the pressure rises, the man goes back into the house and, for the ensuing sunny day, a dainty blond woman rotates out.

Herring joined CoCoRaHS in 2010 and has since collected data from his (low-tech, CoCoRaHS-approved) rain gauge every day at 7:00 a.m. He explains his Jeffersonian morning routine: "Start the coffee, feed the dogs, and check the catch in the rain gauge." When he goes out of town, his pet sitter takes over. As the county coordinator for CoCoRaHS, Herring is responsible for training incoming volunteers in his county. Other than initial trainings, he rarely sees other participants in person. Through the CoCoRaHS website and its newsletter, *The Catch*, participants can view each other's data. During a recent tropical storm in Florida, Herring often viewed data from CoCoRaHs observers down there. Participants know that they are part of a collective effort, even though they operate in physical isolation from each other.

Herring speaks competently and matter-of-factly about his sales job in the home health care industry. In contrast, his eyes are ablaze when he talks about monitoring the weather. He vividly recalls when Hurricane David hit the Virginia Peninsula in 1979, flooding his hometown of Hampton. He witnessed his first tornado when he was eleven. He has seen many waterspouts, which Herring describes as harmless tornados over open water. At his home outside of Richmond, Virginia, Herring enjoys it when he can sit in his den with his son and daughter and "watch the needle drop." He says "watch the needle drop" with a gusto that gives the impression he is a captain watching the ship's barometer, anticipating the need to "batten down the hatches" and face battle with a storm.

Herring is computer savvy and loves his gadgets, but argues that he is "only twenty to thirty percent geek." He shows me his flip phone as evidence: "Look at this, I'm prehistoric." He is a beast when watching storms, hootin' and hollerin' as others might when watching Duke versus North Carolina in a NCAA championship game. Being a spectator of live storms, especially ones with lightning, hail, or strong winds, is not for the faint-hearted. For people like Herring, weather is best as a full-body *experience*. (Though putting safety first means staying inside during storms.)

Experiencing nature with all of one's senses is the heart and soul of science. The Enlightenment philosopher John Locke was a champion of empiricism, the idea that we gain knowledge through experiences of our sensory perceptions. I go out and feel it is hot, use a thermometer to measure how intense the heat really is, and thus I have knowledge that it is hot. Sounds reasonable, almost simplistic. But in the late 1600s competing ideas were more commonly accepted. Philosophers believed that knowledge was innate (we are born with it), or derived from intuition (good hunches), or deduced (figured out by logic).

Experiencing nature with all of one's senses is also the heart and soul of being human. Even though Jefferson wanted weather data to develop a theory of climate and prove North American superiority, he also had an almost spiritual inspiration in his deep-seated love of the seasons. On weather collection, Jefferson wrote, "Climate is one of the sources of the greatest sensual enjoyment."

The scientific insights about storms and terrain as seen via CoCoRaHS data are the cake, and myriad practical applications of the data are the icing. The publically accessible, fine-scale data from CoCoRaHS, all volunteer collected, are used by meteorologists, claims adjusters, attorneys, construction businesses, utility companies, mosquito control experts, farmers, and urban planners, to name a few. In recent years Co-CoRaHS began sharing the data for free, with no strings attached, though that makes it harder for the organization to track all the uses of that data. When people write to say thank you, then CoCoRaHS knows who is using the data and how. For example, CoCoRaHS knows that the Nappanee Missionary Church in northern Indiana uses volunteer-collected data from CoCoRaHS in negotiations when contracting with a company for winter snow removal.

In our market-driven world, the most convincing evidence of the importance of CoCoRaHS data may be the fact that private-sector companies take the publically available data and customize it for clients. For example, engineers and planners developing storm water management plans want gridded color maps that integrate data from multiple sources, including CoCoRaHS and NOAA. Insurance appraisers and roofers work like ambulance chasers, following severe weather events closely in order to be the first of their trade on the subsequent disaster scene. They understand that hailstorms are not uniform across town; they want to know where the weather hit worst so they can get there first.

In Virginia, Herring uses his weather knowledge in his job selling home health care equipment. He doesn't want to deliver a cane or commode and ask, "How are you feeling today?" He knows they feel terrible, because most people who need home health care are ill or have limited mobility. Instead he'll say, "You picked the perfect day to stay home, sunny all day, but tomorrow it will rain buckets."

But talking about the weather *can* be political. "Do friends turn to you for opinions about climate change?" I ask him over lunch. Normally open and straightforward, now Herring hems and haws in discomfort. Instead of using the term *climate change* he notes that we are "putting a curve ball on natural changes," "helping the weather be ornery," or "making angry weather." His politics pose an apparent internal conflict: he has a strong desire to see climate change stopped, but abhors taking away individual liberties. He has similar feelings about evolution, saying he attends church every Sunday, but that he believes in the "Darwin thing" (yes, he also avoids the term *evolution*). Ultimately the complexities of the situation make him feel as though his hands are tied. Yes, Herring reluctantly accepts that human activity is influencing the earth's climate but, no, he does not share his views about climate change with his friends. He knows they would disagree.

Even though he feels helpless, recent research shows that Herring is actually in one of the best positions to make a difference regarding public opinion on climate change *if* he would talk to his friends. The gap between the serious threats from climate change and public policies has prompted a plethora of social science studies to understand how people form beliefs.

Yale University has several researchers, including Dan Kahan, examining how people form beliefs about climate change. Kahan's team revealed that our minds work like a filter, accepting as true those messages from people in our own circle of influence and generally disregarding as false those messages from outsiders, people we don't identify with. Herring's demographic profile is similar to the average demographic profile of climate change deniers geographically, socially, religiously, and politically—you name it, Herring fits the bill.[6] Therefore, if he were to tell friends that he thought humans were causing climate change and that it was urgent to reduce greenhouse gas emissions, his message would likely be heard by his peers; it wouldn't be filtered out.

As far as data goes, it turns out that the ultrahigh variation in precipitation makes CoCoRaHS data not particularly useful for the study of climate change. Many decades—maybe centuries—of data would be needed to decipher trends in local variation. On the other hand, data since 1890 from the Cooperative Weather Observer Network have been the backbone of many a climate change model.

• • •

---

[6] The twenty thousand participants in CoCoRaHS are typical of many citizen science projects: they are older, moderately well-to-do and, as Doesken says, "wildly Caucasian" (Over 90 percent of participants are white).

Thomas Jefferson explicitly acknowledged the rights and responsibilities of citizens in self-governance when he prepared the US Constitution. I wonder how different the United States might be today had he been able to implement a system that explicitly assumed rights and responsibilities of all Americans as citizens of scientific endeavors. Jefferson claimed that a nation can never be both ignorant and free. Freedom, via democracy as he conceived of it, requires an intelligent, informed populace capable and willing to learn. Instead of relying solely on a free press and public education to keep people from ignorance, Jefferson knew that citizens could also be part of science to guide their own education and discovery. Collecting weather data can be a simple civic duty, and the National Weather Service and CoCoRaHS programs are embodiments of Jefferson's dream.

Jefferson may have hopped a stagecoach to share observations with friends, and of course could never have envisioned that, with a few taps on a keyboard, observations could be shared globally and archived in perpetuity. Like the Declaration of Independence, Jefferson's vision for collective science relied on people relishing civic duty and claiming their right to be informed and educated in order to self-govern and curb corruption, privilege, and aristocracy. The United States is becoming a citizen science nation. Whether or not Herring identifies with Jefferson, he, Kientz, and millions like them embody Jefferson's values. To imagine what Jefferson envisioned, people upholding civic duties to leave a data legacy, we only need to see the precipitation gauges populating thousands of backyards across the United States.

In the other chapters of part 1 we will meet more citizen scientists who contribute to the scientific process through their hobbies. Naturalists are hobbyists who are familiar with nature and can identify animal and plant species. Knowledge of natural history is the foundation for all ecological, evolutionary, or conservation research. Despite its central importance, few scientists have natural history expertise. Museums have recognized the decline in professionals with the ability to curate and care for collections and carry out the science of taxonomy. Similarly, there is a decline in such expertise among the public: a wealth of natural history expertise is not getting passed down to up and coming generations. Children can recognize hundreds of corporate logos, but only a handful of local plants and animals. Compare the youth of today with those portrayed in old Hardy Boys books. Kids on iPhones bare little resemblance to the fictional Frank and Joe Hardy, who were outdoors and inquisitive every free hour. Their friend Chet Morton, the chubby farm boy who hid apples in his pockets for snacks, had a myriad of science-related hobbies including spelunking, skin diving, using microscopes, weather monitoring, and infrared photography. Today 20 percent of kids are obese and are occupied in electronic entertainment over seven hours per day. Less familiarity with natural history means less citizen science, and also less conservation. In his 1998 book *The Thunder Tree: Lessons from an Urban Wildland*, Robert Michael Pyle, who in 1974 founded the Xerces Society, a nonprofit organization that works toward the protection of wildlife, warned of the extinction of experience and its consequence for conservation. As he foresaw a future where kids spend more time online than outside, he pondered, "People who care conserve; people who don't know don't care. What is the extinction of the condor to a child who has never seen a wren?"

Ornithology, the topic of chapter 2, is a case in point where the experts in natural history are more often hobbyists rather than professionals. As professional ornithologists focus to become experts in one small area—say, hummingbird tongue and flower coevolution—birders spend their leisure becoming acquainted with the daily and seasonal habits of many bird species. These birdwatchers have enormous uncredentialed skills in identifying birds by sight and sound or by simply looking at the architecture of a nest or the color of eggs.

CHAPTER 2

Ornithology
Bird is a Verb

Education is when you read the fine print; experience is what you get when you don't.

—PETE SEEGER

DAVID WARDEN WAS ABOUT FOUR YEARS OLD WHEN HE WAS LIFTED UP TO see the eggs in a blackbird nest in a garden hedge in Hall Green, a southern suburb of the midland city of Birmingham in the United Kingdom. He was ten when he found a nest all by himself in the same garden hedge. This time it was a dunnock nest. The dunnock—whose name is from the Celtic word *dunn kos*, which means "little brown one"—is a drab bird with a quirky sex life: females mate with many males, over a hundred times per day, which leads each male to first peck at the female cloaca in the hopes of stimulating ejection of sperm from other males, followed by spending one-tenth of a second copulating. For a budding naturalist, it was a bonus that the nest also contained an egg of the cuckoo, a parasitic bird that lays her eggs in the nests of other species. At the time, during the Second World War, Warden had an uncertain future. He scarcely imagined that he'd grow up to be a veterinary surgeon with a small country practice. He absolutely did not imagine that his hobby would include finding over thirteen thousand nests during the next seventy years or that he would still be at it in his eighties. Nor did he know at that time that his sundry observations would be used by researchers half a century later in pivotal publications demonstrating that some impacts of global warming had already arrived. Back then he was simply a kid in a wool newsboy cap and sweater.

33

Warden was born in 1933, the same year that the British Trust for Ornithology was established; five years after finding his first cuckoo-infested nest, Warden began reporting all of his observations of nesting birds to the Hatching and Fledgling Inquiry (which eventually became the Nest Record Scheme) of the Trust. That was 1948, and Warden was unwittingly contributing to our understanding of global warming the year Al Gore—who eventually became a leading climate change activist—was born.

Most commonly, the observations from one individual's hobby does not lead to new discoveries, but new insights emerge from the aggregate of hundreds, thousands, or even tens of thousands of people like Warden. The power of the crowd is a hallmark of citizen science. For example, long-term birdwatchers report that the thrill of finding a nest never fades, "even if it is one's five hundredth blue tit," explains Roger Peart, who has reported bird nests to the British Trust for Ornithology since 1972.[7]

Today the Nest Record Scheme boasts over 1.6 million nest records. Birdwatchers recorded observations of each nest on a nest record card, about the size of an index card. Stockpiled in filing cabinets at the British Trust for Ornithology, many of the cards have been scanned into digital images. In addition, the information on a large subset of over 450,000 cards have been manually entered into a database; one volunteer has entered over 40,000 historic cards alone. The nest records that have been digitized to date have contributed to more than 250 research papers. Many of these papers are only of interest to ornithologists and birders and describe the nitty-gritty natural history of particular species. Two of the most notable papers, however, are about how climate change is affecting birds. These were authored by H. Q. P. Crick and colleagues.

I connect with Humphrey Crick on Twitter; to my gleeful surprise, he has already been following me. Despite his English modesty, I am able to get him to admit that his papers in the highest-ranked journal *Nature*, in 1997 and again in 1999, were "quite groundbreaking at the

---

[7] The blue tit is a cousin to the chickadee and titmouse commonly found visiting bird feeders across North America.

time." I'm not sure if he means groundbreaking in citizen science, ornithology, or climate change, but they were in fact important for all of those. Crick explains that his employer, the British Trust for Ornithology, had started "upping its scientific game" under the leadership of the previous director, the late Raymond O'Connor.

Scholarly papers are the currency in science. Businesses are valued by revenue acquired, but a research project is assessed by how many scientific papers it produces and where these are published. Scientific journals are not all equal. Some are ranked lower, as suitable for small findings, like the basic natural history of where birds nest. Others are ranked higher, for bigger work like testing hypotheses with field observations or experiments. For example, when Crick and David Gibbons used volunteer data to study seasonal changes in the number of eggs laid per nest, Crick explained that "it was quite a coup to get this into *Journal of Animal Ecology*" (an upper-middle ranked journal). A few journals are in the top tier, like *Nature* and *Science*, where the criteria for publication is not only that the study design is sound and the conclusions are supported by the data but that the findings are also of relevance to a broad swath of people. When O'Connor said it was time to deliver the goods, it meant Crick needed to get papers based on volunteer data published in the most prestigious journals possible.

As soon as Crick started looking at long-term patterns in the data of the Nest Record Scheme, "one of the trends that leapt off the page was a large number of species showing trends towards earlier laying." His immediate thought was that earlier laying could be linked to global warming and could be worth a paper. Yet, he says, "A couple of years passed." This is his way, again with English modesty, of saying that he initially failed to realize the importance of preparing such a paper.

At the time it had been established that global warming was caused by human activity, but its consequences to people were still uncertain— let alone its impact on birds. Conservation was place-based. An entire academic field of conservation biology was burgeoning, but the focus was on local issues such as deciding whether a given land concession was better as one large reserve or as a network of small pieces. Upon this small-scale scene arrived a large-scale problem that could not be solved by merely preventing paradise from being paved into a parking lot. It took a while for the significance of global warming research to sink in— even for Crick, who was sitting on valuable climate change data.

Plus, no one really cared about egg laying dates back then. How could early or late laying matter more than clutch size (number of eggs), the number of nestlings, and the overall success of the nest? So even when Crick saw the obvious trends, he tucked the data in the back of his mind, figuring that no one would consider it important.

Crick would have eventually published his paper, but probably not as quickly—and definitely not in the most prestigious journal—had it not been for Chris Mead. Mead worked in the bird ringing program at the British Trust for Ornithology and, in his lifetime, ringed over 400,000 birds of 350 species from eighteen countries. In the worldwide process of ringing (or banding, as it is known in the United States), nonprofessionals dedicate countless hours to helping people like Mead catch birds and place uniquely numbered metal bracelets around their tarsus foot bones (the area above their claws). Birds are banded in the hopes of recapturing them later in life. Those banded as chicks and recaptured as adults provide estimates of survivorship. Those banded during migration and recaptured en route, or on the return route the following year, help researchers decipher migration pathways and the timing of their use. There is a handbook, a sort of avian Cartier catalog, detailing the appropriate size of bracelet for each species to prevent chafing. Professionals and volunteers use specially designed pliers to carefully crimp each bracelet for the perfect fit. Similar to hunting, individuals are required to obtain training and state and federal permits in order to capture and band birds. People who knew Mead make him sound like Hagrid with ringing pliers—he was "a gentle bearded giant," Crick remembers.[8]

---

[8] Mead died at the relatively early age of sixty-two, and thus was outlived by a Manx shearwater that he had ringed in 1957 (the bird was an adult) in Wales. He retrapped it the year he died (2002), and others retrapped it in 2004. The shearwater not only outlived but outtraveled Mead, who estimated that the bird had flown over five million miles in its life.

In April 1997 Mead showed Crick a new paper in *Nature* in which Boston University's Ranga Myneni and his colleagues used satellite data to show that in the northern part of the globe there has been an increase in photosynthesis over recent decades, creating a longer plant growing season, consistent with warmer temperatures. Crick remembers Mead providing him needed encouragement to aim for a brief paper in *Nature* as a follow-up. Mead emphasized what Crick already knew but needed to hear: the British Trust for Ornithology had a stockpile of data for a wide range of species showing unequivocal trends that strongly suggested that this part of the animal world was already being influenced by global warming.

Mead was entirely correct on all counts. *Nature* accepted the paper and published it in August 1997. Crick's findings were something the world needed to hear. The British government, at negotiations to adopt the Kyoto Protocol to the United Nations Framework Convention on Climate Change, referred to the study as evidence that real effects on wildlife were already occurring.

Crick wrote a second paper more tightly linking egg laying dates to climate change. This paper, which he completed with Tim Sparks during six weeks of sabbatical from the British Trust for Ornithology, was also accepted in *Nature* and published in June 1999. He notes, "That was a better paper, as it suggested quite a strong link to climate change that was unlikely to be due to other covarying factors."

Thanks to hundreds of volunteers like Warden who have monitored nests and shared their observations in a systematic way for decades, Crick presented twenty-five-year patterns in twenty very different species that were each tending to lay their eggs earlier and earlier. He also showed over fifty-seven years of observations that the laying date of thirty-one species corresponded to weather—in some cases, temperature, and in other cases, rainfall. For the 1999 paper he analyzed almost a million nest records.

Crick says it didn't seem to matter to anyone that his evidence came from birdwatchers rather than traditional scientists. Many other papers had been criticized for sparse data and so, Crick suspects, "the reviewers were generally bowled over by the sheer amount of data we could present." The global scientific community follows the Royal Society motto *Nullius in verba* to the extreme. Thus, it's never surprising when reviewers raise concerns about the quality of data collected by citizen scientists. As Crick learned, one accepted solution to data quality concerns is in the high volume of observations that can drown out any influence of occasional errors. And so, citizen science puts an addendum on the Royal motto: Take no individual's word for it, but if the masses agree, take their cumulative word. With thousands of observations, researchers even have the freedom to toss the questionable or extreme outliers from analyses and look for patterns and trends in the core. If they detect a signal despite the noise, then the pattern found in the data is trustworthy.

The Americans were not far behind Crick; Peter O. Dunn and David W. Winkler published a paper in 1999 in the *Proceedings of the Royal Society*; using over three thousand nest records submitted by North American birdwatchers, they demonstrated that tree swallows have been incrementally breeding earlier and earlier, resulting in a difference of nine days earlier in 1991 than in 1959, due primarily to increasing surface air temperatures over the years.[9]

If birds don't begin mating at the right time, they can experience what scientists refer to as a trophic mismatch. This means the best food is abundant before the birds need it, as when dessert comes out of the oven before the appetizers are ready to be served. Birds rely heavily on insects as a food source for their offspring, and insects are highly responsive to temperature, many skyrocketing in number when the preferred temperatures are reached. Ornithologists think that birds reproduce most successfully if they time their chick rearing to coincide with a peak in insect abundance. But three developmental stages come before the hatching of chicks: nest building, egg laying, and incubation; thus, birds have to anticipate and plan for the fourth stage in the reproductive effort. They have adaptations, appropriately named biological clocks and calendars, that enable them to anticipate the timing of seasonal changes, assuming conditions are similar

---

[9] The surface air temperatures were also recorded by citizen scientists too; see chapter 1.

year after year. If, however, the insect abundance peaks earlier in spring than the birds are historically accustomed to, thus catching them unprepared, the birds can suffer the consequences. A citizen science study of over one hundred migratory birds that breed in northern Europe has shown that even though birds arrive on their breeding grounds earlier than they did even a decade ago, they don't arrive early enough. Insects, which respond to local temperatures, emerge too soon, and peak before eggs hatch. Birds are better off if their schedule is just right: not too early and not too late. This timing of mating is where natural selection and Goldilocks see eye to eye.

Not surprisingly, the impact of spring occurring earlier is not unique to birds and insects, and is most pronounced in species more sensitive to shifts in temperature. The most well-established responses to global climate change are shifts in the timing of life cycle events: flowers bloom earlier; caterpillars transform into butterflies sooner; frogs call for mates earlier than expected; and people begin sneezing from hay fever at unprecedented times of year. Throughout history, people have made note of various tokens of spring. Like day tally scratches on a prison cell wall, people avoid losing heart by taking note of any sign of winter's retreat. Changes in the signs of spring correspond to two hallmarks of global warming: milder winters and higher spring temperatures.

Dunn and Winkler used nest records of tree swallows from many citizen science projects throughout Canada and the United States, including one of the largest nest monitoring programs in North America, NestWatch. Originally called the Cornell Nest Record Cards, NestWatch was started in the mid-1960s in response to catastrophic declines in some bird species due to the widespread use of the pesticide DDT.[10] Via NestWatch, tree swallows are the second most commonly reported birds after bluebirds. There are three species of bluebirds across the country: the Eastern, the Mountain, and the Western (the range of each of these species correspond roughly to these time zones). It was by

---

[10] The project started in response to Rachel Carson's book *Silent Spring*, which detailed the effects of pesticides like DDT; the book also inspired the annual citizen science project run by the US Geological Service, the Breeding Bird Survey.

studying data on bluebirds in NestWatch when I was employed at the Cornell Lab of Ornithology that I first learned the details of the hobby of nest monitoring.

Finding the nest of a wild bird is not easy, and it is only the first step in nest monitoring. Collecting valuable data requires dedication to visit the nest again and again to determine the pace of its progress and its ultimate fate. Nest monitors are like doctors making their rounds: with each visit they note the health of the nest, anything unusual, and details about the progress of the reproductive attempt—nest construction complete, eggs present, chicks hatched, and so on. Nest record cards in the United States were modeled after the conventions in the United Kingdom, as used by David Warden. A nest record card contains a series of rows, each with a date and details of what was observed at the nest on that date.

While nest monitors take meticulous notes about the same nest over and over again, birdwatchers keep checklists of species seen in a given location at a given date and time. The convention among birdwatchers, which has proven useful for citizen science, is for bird species on checklists to be arranged in taxonomic order. This is a classification system that organizes birds to roughly reflect evolutionary relationships. Such relationships are complex, like family trees, but they are made linear on checklists for organizational purposes. When in taxonomic order, all the ducks are together, all the hawks are together, all the hummingbirds are together, and so forth. A birder might keep a variety of checklists: a cumulative lifetime checklist (commonly called a life list), a home checklist, an office checklist, a travel checklist, a state checklist, a from-my-bathroom-window checklist,[11] and the like. Of course, science is also about listing one's observations. Hence, the perfect match.

Whether nest monitor, bander, backyard birdwatcher, or expert birder, all of these bird hobbyists have contributed to our understanding of birds and climate change, and many other areas of inquiry too. But bird enthusiasts don't always pick and choose which type of bird hobby to have—many do it all, like Kaycee Lichliter.

---

[11] I know people who keep binoculars in the bathroom for such lists.

In 2007, shortly after I began cutting my teeth analyzing NestWatch data for research on bluebird reproduction, Lichliter, a NestWatcher in Virginia who voluntarily managed a trail of nest boxes at the Blandy Farm Experimental Station near Winchester, invited me to give a public talk. Since 2005 she and Greg Baruffi had been participating in a study that I had coordinated about bluebird incubation rhythms.[12] They were both all-around bird enthusiasts, spending their free time monitoring nests, banding birds, and making checklists for a host of different citizen science projects.

Lichliter was drawn into bird hobbies in 2003, under the guidance of several men who taught her to monitor nests and band nestlings. Her most important mentor was Sam Patton, who had been a key part of the bluebird trail at Blandy Farm Experimental Station since at least 1997. Patton took Lichliter under his proverbial wing, immediately spotting her potential to be his successor. On her first day out, a bright and sunny one, she rode around the farm in Patton's car. They would stop on the two-track dirt road, hop out, and hike up to each box. Patton, elderly and shrunken, would remove a latch, open the side wall of the box, and peek in. Lichliter would stand directly behind him and, towering at exactly six feet, peek over his head. No two nests are the same: some have happy outcomes, and some are plagued with disease, parasites, and untimely death. Nest monitors experience both the joys and the harsh realities of nature, one nest at a time. The Blandy trail was handed over to Lichliter the following year, and Patton helped out until he died.

To sustain and grow the trail, Lichliter elicited help. She coordinates a group of about forty-five volunteers each year made up of nurses, retired doctors, lawyers, schoolteachers, and moms and dads. People who don't want to just learn about problems, she told me; they want to get out there and try to do something about them. Getting involved in the day-to-day business of birds and their families can be emotionally

---

[12] They became the poster children for this project, appearing together at a nest box with an incubation data logger in a photo for a news article in *Science* about citizen science.

draining. Nature is harsh, even in our backyards, and nothing is easy once that nest is created: predation, starvation, and hypothermia are not uncommon.

When Lichliter inherited the trail the boxes were on wooden posts without predator guards. Some were occupied by mice, and some had fallen to the ground. After meeting Greg Baruffi during an annual Christmas Bird Count, she recruited him to help refurbish the whole trail with over 130 nest boxes.

The Christmas Bird Count is one of the birding community's oldest citizen science traditions. Its precursor in the 1800s was an annual competitive hunt of birds and small game animals for Christmas feasts. In 1900 the tradition morphed into an event to comprehensively make checklists of living wild birds. Now birdwatchers bundle up against the cold and huddle in groups to make a checklist of birds seen or heard within sectors of their assigned fifteen-mile circle over the course of Christmas Day. There are over two thousand of these preestablished circles for the Christmas Bird Count across the United States.

While nesting data has revealed earlier breeding due to climate change, the winter counts have revealed that species are shifting their winter ranges north. In one study of 254 species recorded regularly in the Christmas Bird Count at various locations around the United States, researchers estimated that these birds have gradually been shifting their winter habitats to the north, by about one kilometer per year on average between 1975 and 2004. Such shifting is partly driven by warmer winters associated with global climate change.

Baruffi was a citizen scientist Renaissance man in his spare time, with expertise in collecting checklists on birds, taking samples for water quality, and helping monitor wildflowers. Lichliter got Baruffi into nest monitoring, and then into bird banding; together they completed a course offered in nearby Warrenton, Virginia, by the California-based Point Reyes Bird Observatory to get their banding licenses. At the nearby Burwell–van Lennep Foundation, they set up a station for the Monitoring Avian Productivity and Survivorship (MAPS) program, a continent-wide effort of mostly volunteer banders who run what is called constant-effort mist nesting stations typically aimed at capturing and banding songbirds.[13] For banding larger birds, Lichliter and Baruffi went to a popular birding spot, Cape May, New Jersey; he helped her get a red-tailed hawk out of a net, and she help him get the hawk's talons out of his hands. Lichliter and Baruffi checked hundreds of bluebird nests and banded hundreds of nestlings. The last thing they did together was their fourth Christmas Bird Count in 2007. Three days later, in the darkest hours of the night, Baruffi's car went into a ditch and he was killed; he was president of the local Audubon chapter at the time.

Lichliter continues as steward of the bluebirds. For people like Lichliter, Baruffi, and Warden, nest watching means keeping an eye on the health of the planet. They are just three of tens of thousands of individuals whose observations in aggregate give us insight into the state of wild birds, and three of millions whose observations give us insight into the state of natural phenomena. Scientists expect some species to be more sensitive to climate change than others, but so far there appears to be no rhyme or reason to the wide range of species affected by it around the world: tree swallow, bluebird, winter wren, dunnock, blackcap, willow warbler, long-tailed tit, greenfinch, and more. These species have a hodgepodge of traits: residents and migrants, insectivores and granivores, single- and multibrooded, great and small. If the planet were a coal mine, then it appears that almost all bird species are canaries.

Avian citizen science boasts a big résumé, with studies of reproductive timing, migration, abundance, and population trends in response to climate change. The bottom line for many birders, however, is the use of citizen science data to document bird distributions and movements in ways that help their conservation. Even though conservation concerns

---

[13] At MAPS stations, mist nets in ten permanent locations are run for six hours, beginning fifteen minutes after the local sunrise. All birds that are caught at the station are identified by species, age, and sex and banded (if not already banded). The ratio of adults to juveniles provides a clue about productivity, and recapture rates of banded birds indicates survivorship.

drove the creation of some iconic projects, like the Christmas Bird Count and NestWatch, the conservation kingpin of bird citizen science was born through the Internet; eBird,[14] a checklist project of the Cornell Lab of Ornithology, began in 2002 and has grown into a global network of birdwatchers sharing checklists with a large interdisciplinary team of researchers, including ecologists, statisticians, mathematicians, computer scientists, climate scientists. The eBird team handles big data, and has published over 120 scientific publications so far.

The team processes data quickly so that it can inform management decisions in a timely way. For example, purchasing land for conservation, a founding premise of the Nature Conservancy, can be an effective strategy for managing wildlife populations. Land purchases of breeding or wintering grounds where birds spend months of their lives can be a cost-effective strategy. But what is the best way to help birds along their migration routes where they use many stopover sites, each for a very short period of time?

The Pacific Flyway is a migration route for shorebirds traveling from the Artic to the southern reaches of South America, and it cuts through central California. The fertility of California's Central Valley stems from the network of rivers, creeks, and sloughs that for millennia spread nutrients like repeated coats of paint with every flood of the surrounding land. More than 95 percent of the original wetlands, a natural stopover site for migrating water birds, have been lost, mostly converted to fields of almonds, apricots, asparagus, and avocados, and on

---

[14] Another Cornell Lab checklist program is Project FeederWatch, which began in Canada with the Ontario Bird Feeder Survey in 1976. The project eventually spread across North America after becoming jointly administered with the Cornell Lab in the 1990s. Project FeederWatch engages tens of thousands of North Americans in gathering checklists of birds that visit feeders filled with millet, sunflower seeds, suet, nectar, corn, and more. This winter-only program involves a unique type of checklist: participants report the highest number of each species of birds that visit their feeders each week. For example, if a person sees two black-capped chickadees one day and three later in the week, they report three (not five) on their checklist. These collective observations have provided insights into bird community structure, competition among bird species, and patterns of the spread of the nonnative Eurasian collared dove.

through the alphabet to zucchini. Despite the extreme losses, the area continues to host the highest density of migrating waterfowl in the world—with peaks of seven million ducks. Still standing after one punch of wetland loss, the ecosystem could be knocked down with a second punch.

The second punch has arrived in the form of years of extreme drought in the region, which has left millions of migrating shorebirds and waterfowl with even fewer stopover sites. The Central Valley supports 30 percent of shorebirds and 60 percent of waterfowl that use the Pacific Flyway. In this situation, the Nature Conservancy decided that it could help birds on migration simply by renting rather than purchasing land with the right habitat; it created a market-based program called Bird Returns, which has enrolled over forty farmers along the California portion of the Pacific Flyway. Through reverse auction, every farmer submits a bid, and the Nature Conservancy choses whom to pay to temporarily flood their fields with two to four inches of water during spring and fall migration. More than ten thousand acres are flooded on four-, six-, or eight-week contracts, both leading into and following migration. The key to the project's success is identifying the right acres to flood within the enormous Central Valley at exactly the right time.

The Nature Conservancy uses eBird data to decide which "pop-up" wetlands to fund during northward and again during southward migration. Citizen science eBird volunteers in California have submitted over 230,000 checklists of the area. With the eBird research team using high-performance computing, they forecast where birds are likely to be present and overlay NASA images of surface water. They spot mismatches and advise the selection of farmers in locations to literally fill the gaps with water. During the initial test, birds used the pop-up wetlands in densities at least twenty times higher than nonflooded neighboring fields. Birders celebrated sightings such as twenty thousand dunlins at a time. All fifty-seven species of shorebirds that migrate through the Central Valley used the pop-up wetlands, for a total of 220,000 birds recorded in pop-up wetlands during migration.

The eBird program's citizen science adjusts quickly to conservation emergencies too. In April 2011, and for the following five months, the BP Deepwater Horizon oil disaster became the largest oil spill in history, releasing more than 170 million gallons of oil into the Gulf of Mexico. The diverse ecosystems of the gulf waters, the adjacent coast, and the Mississippi River Delta were affected, and the long-term impact is still coming to light. In the immediate wake of the disaster, as people on shore knew the oil slick was approaching barrier islands, estuaries, salt marshes, and other coastal wetlands, the eBird team rapidly modified their system for data collection and display in the hopes of aggregating information to help save birds. The team decided to focus checklists on ten species of known conservation concern, including the roseate spoonbill, the American oystercatcher, and the brown pelican, because these species had only recently been removed from the endangered species list.

At the time, there was a media blackout and people desperately wanted more information. For example, a group formed to float balloons holding cameras over the coast in order to get high-resolution aerial photographs to track the progress of the oil slicks. Initially called Grassroots Mapping, the group later merged with the Public Lab, a ragtag global community in which people share do-it-yourself techniques for monitoring environmental conditions.

In this data void, the creation of the Gulf Coast Oil Spill Tracker as a module within eBird was a valuable source of timely information. The tracker was a mash-up of several data sets in a real-time map of five Gulf Coast states, continually updating the display of all submitted records on the locations of the ten bird species of conservation concern, the current extent of the oil slick, and the seventy-two-hour forecast of the oil slick. The data visualization tool relied on real-time data from eBird participants, Google Maps, oil slick data, and a forecast provided by the Satellite and Information Service, a division of the National Oceanic and Atmospheric Administration (NOAA).

The eBird program also modified its web data-entry system to allow observers to report specifically on oiled and sick birds. With a simple question, "Would you like to provide comments or more details about a species (e.g., if a bird is oiled, age/sex, etc.)," it quickly received about nine hundred observations of oiled birds.

The oiled bird reports and the combined maps were useful to steer rescue efforts. Beach cleanup and bird rescue crews needed to make triage decisions to focus their efforts on the species most in need of help and most likely to benefit from rescue. As Brian Sullivan, one of the eBird project managers, explains, "If we get a report of fifty Brown Pelicans with light oiling on one bird, that's not so bad. But if we get a report of ten Brown Pelicans, and every single one is coated with oil, that's an area getting badly hit, and it should be a priority for cleanup and bird conservation work."

The wealth of checklists already collected by the eBird program provided baseline data about bird communities before the spill. In the year following the disaster, more than four thousand birdwatchers along the gulf submitted over 110,000 checklists. Tallies of birds found dead from oil exposure underestimate the true number affected because many bodies are not found. Researchers are using eBird sightings to estimate the likelihood of detecting a bird killed by the oil, and based on those detection estimates they will be able to more reliably estimate the number of birds that died but were never found. A lesson from eBird is the high value of baseline monitoring from leisure birdwatchers, which is too often recognized as priceless only when something unanticipated and tragic happens.

Globally there are over 270,000 eBird participants who collectively have had their eyes behind binoculars for over twenty million hours to provide over 280 million bird observations. It's worth noting that not all participation is the same. About half of the time, people submit checklists that give a complete account not just of the birds present in an area but also those absent from the area.[15] Most of the bird enthusiasts who use eBird simply enjoy its free and plentiful information,

---

[15] After filling in a checklist, participants are asked, "Are you submitting a **complete checklist** of the birds you were able to identify?" If, for example, someone is reporting a "best of" checklist, or excluding nonnative species, or only reporting hawks, then they answer no.

often using it to plan their birding trips. Some organizations and federal agencies have mandates to share data they collect, but do so as downloads of cumbersome raw data tables. The eBird program provides data access through state-of-the-art visualizations so that people can make sense of the data. Each year millions visit the eBird website and view dynamic maps of recent observations and other intuitive graphics. This means that a small portion of eBird users, perhaps as few as 1 percent of them, are responsible for submitting the vast majority of the bird sightings. But more people using the data than contributing data doesn't hamper the program.

The inequity in effort among eBird users is common in citizen science, as it is with other types of crowdsourcing and online engagement in general. For example, most people on the planet with Internet access use Wikipedia, but according to Jimmy Wales, one of the founders of Wikipedia, only 9 percent of Wikipedia users edit existing content and just 1 percent create new content. This pattern is known as the 90–9–1 principle. Wales asserts that a core group of contributors do virtually all the work, with the help of occasional contributors who do minor edits. According to this view, most of us take a free ride.

Aaron Swartz, who achieved fame in his teen years for helping develop the web feed format called RSS, has a different take. While Wales created his estimates by counting edits to Wikipedia, Swartz counted the total number of characters added to Wikipedia by each person, which led him to the opposite conclusion. By Swartz's summation, tens of thousands of contributors pitched in a little content to Wikipedia and a core of five hundred to a thousand regulars carried out edits such as correcting spelling and formatting. Swartz's view is that "the formatters aid the contributors, not the other way around." According to this view, each casual contributor tosses in a bit here and there, and cumulatively these people create the bulk of Wikipedia. Certainly it makes sense that an encyclopedia could be filled quickly by many users rather than a small group of know-it-alls. With hard-core birders, however, maybe the core can see it all.

In eBird's early years, the program used the tagline "Birding for a Purpose." This captured some altruistic birders, but not a sufficient number. In 2006 the eBird project managers changed their strategy and eventually adopted the slogan "Birding in the 21st Century." The program showed birders what's possible with their collective observations if they record the number of individuals of each species seen rather than simply a list of species seen. Thus hard-core birders wanted to embrace the maturation of their hobby as it morphed into a technique that impacts science and conservation.

Whether banding, monitoring, or making checklists, all birders like to notice things. Consider a tree that holds a huge and rambunctious flock of cedar waxwings who have unwittingly become rip-roaring drunk. When fruit sits on bushes all winter, it can ferment, turning each holly berry into a bright red shot of liquor. Consequently, cedar waxwings often become inebriated in the spring. Four out of five people who pass a tree filled with waxwings will only notice the tree; one in five, who may or may not call themselves birdwatchers, will notice a very dignified-looking flock of birds behaving in a relatively undignified way. The ones who notice, who claim extra moments to enjoy and who take note of their surroundings and/or add the birds to their checklists, are the ones rewarded with the peculiar dramas of nature. And those who take the added time to share their observations reward all of us with a world of birds.

If all citizen science required a high level of expertise and the ability to identify drunken waxwings, then the skew in participation would be even greater. But much of citizen science simply requires time. For many of us time seems a rarity, but for others time is the only thing on our side. W. C. Minor was a deranged murderer who contributed more words than any other to the crowdsourcing effort that created the Oxford English Dictionary. He had time to make these contributions during his confinement in a London insane asylum from the late 1800s until the early 1900s. His contributions ended when his dementia worsened and he cut off his own penis (the word for it is autopeotomy) with a knife he had access to because of his dictionary contributions. As we'll see in the next chapter, there are all sorts of people with time to be involved in citizen science, even inmates, each enjoying the sense that they contribute to a larger, collaborative effort.

CHAPTER 3

Entomology
Of Monarchs and Men

Life is the first gift, love is the second, and understanding the third.

—MARGE PIERCY

ON A SEPTEMBER MORNING, SEVERAL INMATES GATHER WITH BILL Coleman, a correctional mental health counselor at the Washington State Penitentiary in Walla Walla. Men convicted of rape, murder, and armed robbery cast their eyes toward the horizon as their hand-raised and tagged monarch butterflies escape over the fence. Each monarch tilts and staggers away in uneven bursts of flight—up and down, right and left—as if intoxicated by the prospect of freedom.

The program of raising, tagging, and releasing monarchs by incarcerated citizen scientists began in 2012 in eastern Washington.[16] The project follows in the footsteps (or shall we say flight paths) of citizen science to tag monarchs in the population east of the Rocky Mountains, which began in the 1940s. Then there was no need to raise and release monarchs for tagging other than for educational purposes; like many insects, monarchs occurred at such high densities that, with a butterfly net, individuals could catch and tag a hundred of them on a short jaunt through a meadow, have a picnic, and then catch and tag a hundred more. These epic volunteer endeavors in the east led, decades later, to

---

[16] In Washington, citizen science with inmates started in 2008 with the Sustainability in Prisons Project, a partnership between Evergreen State College and the Washington State Department of Corrections, when prison staff, inmates, and scientists began teaming up to restore endangered species and habitats. The program also promotes sustainable prison operations through energy conservation, recycling, and other practices.

solving one of the biggest mysteries in the field of entomology: Where do monarch butterflies throughout the eastern United States and Canada go during winter? (Answer: Mexico.) The western efforts may also take decades to resolve their mystery: determining whether monarchs born west of the Rockies all migrate to California or whether some migrate to Mexico like their eastern counterparts. This particular raising and tagging program could provide conservation benefits for what might, sadly, be the next endangered species.

When I ask butterfly researcher Leslie Ries how important citizen science is to solving the monarch mystery, she responds by simply noting a momentous date: "January 9, 1976." Ries leads an effort to unify North American data from the multitude of programs through which people have reported their butterfly observations. There have been continental, regional, and local butterfly monitoring programs with volunteers for decades. There are programs based on sightings of adults, counting caterpillars, carrying out standardized counts of adults and caterpillars, systematic atlases, transect surveys, and tagging programs. All include valuable observations, which Ries is making more valuable by bringing them all together and making them available to researchers and the public.

The numerous programs with adult monarchs and caterpillars are related like cousins and, however indirectly, were inspired by the achievement on that momentous date in early 1976. On that day, in the remote Trans-Mexico Volcanic Belt in central Mexico, Fred Urquhart discovered a monarch butterfly that had been tagged by his volunteers two thousand miles to the north in Chaska, Minnesota. The volunteers were Jim Gilbert, a schoolteacher, and his middle school students Jim Street and Dean Boen. They had placed sticker PS397 on one of the wings of the butterfly in August 1975. For the first time Urquhart had indisputable evidence that monarchs in Canada and the United States migrate all the way to Mexico in the winter.

Urquhart's quest to know where monarchs spent the winter began with childhood curiosity. His wife Norah Roden Urquhart had also wondered as a child where her monarchs went when they made a mass exodus every autumn. Like these two Canadians, Mexican-born Catalina Aguado enjoyed monarchs—perhaps, as it turns out, the very same ones—when she was a young girl. I don't know whether the young Aguado wondered where her monarchs went every spring, but as an adult, she was crucial to the success of the Urquharts' quest.

The pursuit began in 1940 when the Urquharts, after much experimentation, figured out a way to tag monarchs without harming them.[17] They devised small stickers with numbers that also read "Send to Zoology University of Toronto Canada" and created the Insect Migration Association, which would eventually enlist thousands of volunteers in tagging hundreds of thousands of monarchs. Today the small organization Monarch Watch continues the tradition of volunteers tagging monarchs. All volunteers are warned not to wear sunscreen or bug spray so as not to harm their study subjects.[18] After catching a monarch in an insect net, a lucky volunteer holds the monarch's body and lets the wings open to check for the presence of a black spot on the lower half of each wing. This a means of sexing: the spots are pheromone sacs, and they are only present on males. The volunteer then gently closes the butterfly's wings and places a sticker on a particular part of the wing called the discal cell. (On monarchs, the wing veins are black and the areas in between, called cells, are orange, except along the edges of the wings, which are black with white polka dots; the large and oblong center orange cell at the wing base is the discal cell.) The volunteer records the tag number, butterfly sex, and condition, as well as the location, date, and time of the capture and release. For forty years the Urquharts organized citizen scientists across the United States and Canada in this sort of tagging of monarchs.

What did recruitment into citizen science look like in an era prior to the Internet and social media? To recruit and retain volunteers, Norah Urquhart shared tagging information in a periodic newsletter.

---

[17] In the early 1900s, adhesive labels had to be licked to activate stickiness. In 1935, Stan Avery invented self-adhesive labels with peel-off backing

[18] Mosquito repellent is toxic to a wide range of insects, and sunscreen inhibits respiration by blocking insects' breathing tubes, called spiracles.

To recruit, she posted a full-page advertisement in the *Minnesota Citizen* calling for volunteers to investigate the migration of monarch butterflies. In 1952 she wrote a magazine article titled "Marked Monarchs," asking for more volunteers. Twelve people responded to the first call. By 1972, the International Migration Association had six hundred volunteers and thousands of others helping tag. At this time, Norah Urquhart upped her efforts and wrote to newspapers in Mexico, calling for volunteers to be on the lookout for tagged monarchs. Today, for some rural Mexicans, the monarchs are like satchels of money gliding over the border. When scientists from north of the border visit, locals can get as much as twenty US dollars for a butterfly with a tag. Back in 1972, Ken Brugger saw one of Urquhart's ads while he was on a quest of his own in Mexico. Brugger was trying to locate a young woman he had met and fallen in love with during a school trip. (That's what courting looked like in the days before the Internet and social media.) The young woman, who did eventually become his wife, was Aguado.

Monarchs flew through Aguado's part of the Mexican woods every winter. She knew the region and, unlike the color-blind Brugger, she could see the full spectrum of colors. She could also speak Spanish and the local dialect. Aguado and Brugger spent two years exploring her native region until they found a secluded winter roost for millions of monarchs in the Cerro Pelon Mountains west of Mexico City. This was 1975, and they spent that winter looking for butterflies with tags and came up empty. The following winter Fred Urquhart arranged to visit, along with a photographer from *National Geographic*. Within five minutes of his arrival, Fred Urquhart looked at a tree branch festooned with monarchs; the branch happened to break, and in the fallen bounty he saw one with a tag. *National Geographic* captured the moment, and Aguado appeared on the cover of *National Geographic* in August 1976, festooned with monarchs.[19] Eventually Fred and

[19] Sue Halpern, author of *Four Wings and a Prayer: Caught in the Mystery of the Monarch Butterfly* (2001), chronicles the disagreements about scientific credit that resulted in Aguado dropping out of the Monarch story for decades.

Nora Urquhart were awarded the Order of Canada in honor of orchestrating the citizen science work and for a discovery that apparently satisfied what I suspect was a very common, though often unspoken, curiosity.[20]

Around the time the Monarch migration mystery was cracked in North America, far away in Manchester, England, David James had recently completed his Bachelor of Science degree at the University of Salford and would soon be making his way to carry out graduate research on monarchs on the outskirts of Sydney, Australia. Monarchs can be found all over the world, but long distance migrations were thought to be unique to those in North America. Through his graduate research James discovered that monarchs in Australia migrate distances of two or three hundred miles. He dreamed of studying the North American migrations one day. Life's twists and turns led him to settle in eastern Washington in 1999, working as an entomologist focused on integrated pest management—the use of biological agents, like natural insect predators, to control agricultural pests and thereby reduce pesticide use. James focuses on biological agents suitable for "irrigated horticultural cropping systems" in Washington, and such irrigation is most often used on hops production for beer and grapes for wine.

Decades of tagging programs had led entomologists to a good understanding of monarch migration for populations in eastern and central North America. Monarchs are roughly divided by the Rocky Mountains and western monarchs are thought to mostly overwinter in groves along the California coast. There is less data, however, on movements of monarchs west of the Rockies, and there has been no large-scale tagging done on butterflies in the Pacific Northwest, but with good reason: monarchs have always been sparser in the Pacific Northwest, and, given the low odds of resighting, thousands of monarchs must be tagged in hopes of resighting just one. Consequently, the only feasible tagging program would have to involve mass rearing in

---

[20] They received the award in 1998. Fred died in 2002, and Norah in 2009; both were ninety years old.

captivity, followed by tagging and release. With no data, then, people continue to assume that monarchs in the Pacific Northwest are the ones that spend the winter in California.[21] But James wanted to know for sure.

Unfortunately, James was too busy at his day job with predatory insects to be the proper caretaker of a monarch nursery. His very best effort produced about fifty adult monarchs for every hundred eggs. That's not too shabby: in the wild, the most optimistic estimates are that seven out of a hundred eggs become adults, because monarch eggs and newly hatched caterpillars are prey to ants, spiders, wasps, beetles, and more, but fifty tagged monarchs would not be enough. So James teamed up with Tamara Russell, a psychologist and the clinical director at the Washington State Penitentiary, to entice prison inmates to raise monarchs. The Butterfly Wranglers, as they call themselves, have proven to be dedicated individuals, highly capable of rearing butterflies. In fact, they are better at growing monarchs than James, the professional: they average about eighty adults for every hundred eggs, and can accommodate enough eggs to produce two to three thousand adults per year. James is pleased with their abilities. "It is cliché to say it, but they have time on their hands," he tells me. "They can rear them with great care." While James's research benefits from the collaboration with Walla Walla inmates, it is a win-win situation because the inmate experience is therapeutic. No one disputes that the program calms the Butterfly Wranglers and keeps them focused. Participation is entirely voluntary, and inmates are screened for the program based on multiple criteria, which includes their crime(s), behavior while in prison, mental health status, and any past instances of cruelty to animals.

In US prisons, inmates are grouped into five levels based on the risks they pose. The second-highest-risk inmates are classified as requiring close custody. All of the Butterfly Wranglers are not only in close custody but are classified as needing mental health and/or protective

---

[21] Tagging of monarchs in Arizona revealed that some migrate west to California and some south to Mexico.

custody, which means they are isolated from other inmates who might harm them in reprisal for the nature of their crimes. The Butterfly Wranglers may appear to be double castaways, separated from society and other inmates, yet while their empathy toward fellow humans has suffered, they are often hesitant to enter the monarch program because they are afraid they will accidently hurt the caterpillars, thus demonstrating great empathy toward creatures quite unlike us.

Other than having symmetrical bodies and complex societies, humans have little in common with insects.[22] In addition to their abundance of legs and wings, there is much about insects that is inverted: their skeletons are on the outside of their body, and their mouthparts are external. Their ears can take several different forms: some insects, like crickets, have ear-like tympanal organs; others, like mosquitos, hear with an organ located on the second segment of their antennae; and others still, like butterflies, hear with body hairs called setae.[23] Hawk moths (which look like tiny hummingbirds) have a pilifer, an organ in their mouthparts that allows them to hear the echolocation calls of bats. Insects have aspects that to us would seem to be those of superheroes: organs to detect infrared color, X-ray radiation, and the earth's magnetic field. It's no wonder that insects inspire the design of alien characters in B-grade science fiction movies.

Strangest of all, insects look completely different as they get older. We talk about stages of life when comparing our college days to our midlife careers, but we essentially look the same as we age, just grayer and droopier. Insects reinvent themselves entirely from one life stage to

---

[22] Many children are able to identify with these undersized neighbors. Insects are small and yet gain attention for their outrageous colors and forms; they are sometimes beautiful and sometimes fascinatingly hideous. No less important, insects provide ecological services like pollination, pest control, the decomposition of waste, and serving as meals for many other animals—particularly birds. The affinity that kids have toward insects has sparked citizen science projects aimed at gathering information on those insects within the reach of children. This include ladybugs, ants, fireflies, damselflies, dragonflies, butterflies, and moths. Schoolchildren and youth groups have even been involved in citizen science on dung beetles and cockroaches.
[23] Setae have multiple functions, including touch, sensing temperature, and even sensing humidity.

another, turning into something completely unrecognizable compared to what they once had been. (Hence the term *metamorphosis*.)

Monarchs have four stages to their life cycle, which is fairly run-of-the-mill for insects. They begin as an egg, hatch into a caterpillar, molt into a pupa, and metamorphose into an adult.

What makes monarch migration notoriously mysterious (other than the now solved question of *where* they migrate to) is that they partition their life cycle into four generations per year, each with a purpose.[24] This means the butterflies that migrate south are the great-great-grand-children of the adults that migrated north. Young birds will migrate south without ever having been there before, but they may follow older birds who have made the journey before. Migrating butterflies, on the other hand, are entirely a pack of youth with no experienced leaders to follow; the fourth generation simply knows instinctively what to do. According to the legend in the former television series *Buffy the Vampire Slayer*, "Into every generation a slayer is born." The monarch legend is thus, "Into every *fourth* generation a migrator is born."

Let's burrow into the bizarre pattern that skips over three genera-tions, which is well understood for the populations that spend the summer east of the Rockies. Monarch adults leave Mexico in February and March, migrate north, and breed.[25] Most die soon after they lay eggs in Texas and other southern states, and this is generation 1. Those eggs go through the four stages of life, eventually emerging as the generation 1 butterflies, who migrate northward and enjoy life for two to six weeks, laying the eggs of generation 2 before they die.[26] The eggs of generation 2 develop in May and June, and those adults

---

[24] The number of monarch generations per season varies geographically. In Washington state, there are only three summer generations. In much of the US Midwest and Canada, there are four generations.

[25] The animated maps of annual northward expansions and southern migrations from a classroom-based citizen science project called Journey North give excellent examples.

[26] Butterfly lifespans are at the whim of the weather. Monarchs live longer when days are cloudy; A couple of sunny weeks and they are buttered toast. It makes estimating population trends difficult because you can't readily tell whether a declining trend is representative of the population total or individual life spans.

lay the eggs for generation 3 in July or August. Generation 3 eggs reach adulthood in time to lay generation 4 in September and October. These eggs start out just like generations 1, 2, and 3, but when they become adults they opt against the James Dean path of living fast and dying young. Instead of rushing into young love, they migrate south and live in Mexico for six to eight months, and migrate back the following year to mate and lay eggs for a new generation 1. Thus, an equally accurate monarch legend is, "Into every fourth generation an *abstainer* is born."

In Washington the season is shorter and monarchs have three generations per season. James waits until mid- to late summer, when he's most likely to get the second generation, to collect a few female monarchs and bring them to his lab where they lay eggs of the third generation. "A single female will lay up to five hundred to seven hundred eggs," he told me.[27] He then delivers the eggs, which are typically on the underside of leaves on milkweed plants, to the prison in Walla Walla.

The eggs hatch within three to eight days. After hatching, the inmates keep the lab temperature at seventy-four degrees, which helps the caterpillars stay active. "They want to eat as much as they can, as fast as they can," explains Bill Coleman. Fresh food is essential because caterpillars grow rapidly, and they are finicky children, consuming only one item: milkweed. The inmates are not allowed to leave the facility to gather fresh milkweed, and a plan to grow it in the prison vegetable garden was nixed.[28] Consequently, Coleman and other staff gather fresh milkweed almost every day, and grow some outside the prison walls in case of urgent need. The caterpillars grow so fast that they outgrow their skin often and shed it, much like a snake does; the stage after each shedding is called an instar, and monarch caterpillars develop through five instars. The first instar is pale green and translucent. The second one

---

[27] There are monarch farms in California that sell eggs, but James stresses the importance of native stock.

[28] The concern was that milkweed could be used as a poison. It contains digitalis, which stimulates heart activity. If a person eats a high amount of it, he or she could have a heart attack.

gains yellow, white, and black stripes. By the fifth instar, at nine to four-teen days of age, the caterpillar is two thousand times larger than at the first instar. Even though caterpillars' intentions are to eat only milkweed, they will chomp through whatever is on the milkweed, including monarch eggs and smaller instars. To prevent butterfly cannibalism, the inmates check every day to make sure the milkweed is fresh and plenti-ful, and segregate the instars by size. When it is time for a fresh plant, they painstakingly move each caterpillar, sometimes with tweezers, to the new plant and put them in a newly cleaned holding barrel.

Before a caterpillar's final molt it produces silk to fasten its body to a leaf or twig. Then the outer layer of skin peels away and the animal enters the pupal stage, that of the chrysalis. They emerge from the chrysalis as adult butterflies ten to fourteen days later with small, wet wings and weigh as little as a paperclip. The inmates wait twenty-four hours for the wings to fully dry, turn down the temperature because chilly butterflies are easier to handle, and then capture and tag the young monarchs. Even though James delivers all the eggs at once, they don't all develop at the same rate or emerge at the same time. When fifty to sixty adults are emerging per day, prison staff will take the tagged insects to flyways where they will have a strong chance of find-ing milkweed after their release. As the season dwindles, and just a few adult butterflies are ready at once, the inmates will let them go from within the prison yard.

The rearing process only takes three to four weeks, but it can take years, decades, or even a lifetime for tagging efforts to yield valuable data. Walla Walla has had many local resightings, but only eight re-sightings of long-distance movements thus far, which is not enough to as yet derive any conclusions. One of the prison's launches made it to Brigham City, Utah, which doesn't exactly seem like the best path to California, but others were found in California—the farthest near Santa Barbara, more than eight hundred miles away. Resightings are reported via cell phone pictures.

James claims that butterflies have universal appeal and that he has witnessed children with bright futures and hardened felons with the

poorest of life's prospects respond identically to monarchs: with innocent awe. He notes that, without exception, "these murderers and real
criminals were really gentle with butterflies. They didn't want to harm
butterflies. They really didn't want to hurt them even though they had
done much worst in the past."

Karen Oberhauser, an entomologist at the University of Minnesota,
runs the Monarch Larva Monitoring Program (MLMP), through which
volunteers count and measure larvae on milkweed plants in the wild.
Over one-third of the adult participants engage kids in the work outside
of formal school settings—via 4-H Clubs and summer camps, for example. The adults note that kids love using field equipment—cruising
butterfly habitats with insect nets and hand lenses in the great outdoors;
only 2 percent of the kids most enjoy taking eggs home to watch metamorphosis (which is what the Washington inmates do); a whopping 98
percent most enjoy finding eggs and caterpillars on plants in the wild.
Freedom rules!

So that I could learn as much as possible about monarch butterflies,
I attended an academic seminar by entomologist Ries, who had noted
the momentous date that Fred Urquhart found a tagged monarch in
Mexico. She was visiting from the University of Maryland to speak at
the Cornell Lab of Ornithology. Given the venue, Ries was wise to start
her talk by pointing out the similarities between birds and butterflies.
She explained that ornithologists and entomologists undertake "the science of pretty things with wings." Birds and butterflies are diverse
groups, with ten thousand species of birds and twenty thousand species
of butterflies worldwide. Honing in on the United States and Canada,
the scales tip toward birds, with about 900 species of birds and about
750 species of butterflies. The species have broad mixes of traits: some
migrate and some ("residents") do not; some feed on a variety of plants
("generalists") and some feed on only one type ("specialists," like monarchs on milkweed); many eat plants, but some are parasites (living off
other insects), and the caterpillars of some butterflies are carnivorous.
The biggest similarity is that people are drawn to watching both groups;
the biggest difference is that among butterflies, there is one flagship

species that everyone knows: the monarch. In the United States, about thirty-six thousand volunteer hours are dedicated to studying monarchs each year. About the same amount of time goes into studies of the other 749 species combined.

Butterflies and birds that migrate during the day have one more thing in common: they use their innate ability to tell time—their biological clocks—not only for knowing *when* to fly south but also for knowing *how* to fly south. Since the sun crosses the sky as the day progresses, any daytime migrant, bird or bug, needs to know the time of day in order to properly adjust its bearings. This is called time-compensated sun compass navigation.[29] With this type of navigation a butterfly turns like a dial on a compass based on the time of day and the position of the sun. To be clear, these animals don't use the sun to pinpoint the time of day. They need to use their internal sense of time, compare it to the sun's location, and adjust their orientation, much in the manner that navigation by an eighteenth-century sailor involved a timepiece and a sextant.

In 1927 Austrian Karl von Frisch published *Aus dem Leben der Bienen* (The dancing bees), which explained his studies showing time compensation in bees. Once bees find a nectar source and want to share that location with the workers in the hive, they do a little dance that delineates directions to that source. The directions only work in conjunction with the bees' internal sense of time (*Zeitgedachnis*), which was demonstrated as central to bee navigation: When scientists turned the lights in a beehive six hours ahead of the actual sunrise, bees reset their internal clock to this "false" sunrise. Then, when allowed to leave the hive at noon on a clear day, these bees mistakenly viewed the noon sun as the 6:00 p.m. sun, which one would expect to see in the west. So, if the bees wanted to fly south, they would turn ninety degrees from the sun and mistakenly fly east. With a six-hour shift in timing corresponding to ninety degrees of the twenty-four-hour 360-degree day, bees from a hive with a six-hour shift would consistently turn right in error, making a

---

[29] Depending on the species, birds also orient via magnetic compass, landmarks, and the stars. Put antsy birds in a planetarium, and they will settle down if you project the appearance of the night sky at their destination!

ninety-degree mistake. The fourth generation of monarchs likely have the same time-compensated compass that was found in bees.

West of the Rockies, monarch populations have declined by 50 percent in the last twenty years, and the cause is probably due to a decrease in milkweed in the western breeding areas. While zoos and aquariums have captive breeding programs for some critically endangered species, the goal of the Washington inmate program (which has contributed well over ten thousand monarchs to the Pacific Northwest population over the past three years) is simply to better understand migration; it is not to boost monarch populations. To the contrary, the researchers and monarch conservationists that I interviewed, including James, frown upon large-scale captive breeding and releasing programs for monarchs because of several potential negative impacts. Problematic programs are those for commercial purposes where monarchs are bred for many generations and at high densities. Commercial practices can spread disease among monarchs and reduce the genetic diversity of the species. James emphasizes that "wedding-released monarchs are not going to help monarch conservation, and in fact will likely hinder it." He designed the program with inmates carefully to prevent the spread of disease and loss of genetic diversity; he catches females fertilized in the wild to ensure genetic diversity and screens these wild-caught female monarchs for disease before bringing their eggs in for rearing. Plus, the rearing is interrupted by at least ten months of no rearing, and winter cold weather, both of which would prevent any disease that might make it in one year from carrying over to the next. The prison programs amplifies the success of naturally occurring monarchs, and James emphasizes that the proof is in the pudding: "The very fact that we are able to recover so many reared individuals months after release is evidence of the health of our reared butterflies and evidence of our contribution to the conservation of the species."

East of the Rockies, monarch populations have declined by more than 90 percent in the last twenty years. That's an alarming statistic.

Monarchs are collateral damage in several battles being waged across North America. One of the primary causes of declines in monarch populations has been attributed to the use of Monsanto's herbicide Roundup, which has enabled the mass destruction of milkweed in or near commercial crop fields. There are over one hundred species of milkweed native to North America, and the core of that distribution has always been the Midwest prairies. Even after many prairies were converted to intensive agricultural fields, milkweed persisted in high densities among the corn and soybean crops.

Monarch caterpillars subsist on only about twenty-five species of milkweed, and have preferences for a select few.[30] Their dependence solely on milkweed makes them exceedingly vulnerable to changes in agricultural practices. But milkweed is considered by many to be a weed, as its name implies. As a result, Monsanto devised the herbicide glyphosate, most commonly sold under the name Roundup, which kills plants by blocking a particular enzyme (EPSPS) that synthesizes essential items like vitamins. Some soil bacteria have a form of EPSPS that does not get blocked by Roundup, and other bacteria can produce an enzyme that degrades the blocking agent in Roundup. With genes from either bacteria, resulting genetically modified (GM) plants are resistant to the herbicide.[31] GM crops like Monsanto's Roundup Ready corn and soybeans came to the agricultural market only in the late 1990s, but today they already make up 94 percent of all soybeans and 89 percent of all corn grown in United States. The resistant plants have led to a twentyfold increase in the use of Roundup, now to the tune of 204 million pounds per year.

Milkweed is a regular plant, with no protective bacterial genes in its cells, and the heavy use of Roundup has decimated the plant in the

---

[30] Adult monarch butterflies prefer other nectar sources, though bees like milkweed nectar.

[31] Genetic modification (or genetic engineering), also called recombinant DNA technology, involves transplanting the genes that express the desired trait of the donor organism (in this case, soil bacteria) into the recipient (corn) so it can also produce the same trait as the donor (either an enzyme that degrades the blocking agent in Roundup or the EPSPS that is not blocked by Roundup).

Midwest, which is the heart of the monarch range. One study, by John Pleasant at Iowa State University and Karen Oberhauser of the University of Minnesota and the MLMP, found that the steady loss of milkweed in the Midwest is tightly correlated with the steady decline in the size of monarch groups overwintering in Mexico. A subsequent study led by Tyler Flockhart at the University of Guelph found that the herbicide practices associated with GM corn was the primary factor causing monarch declines. Flockhart and his colleagues wrote, "Recent population declines stem from reduction in milkweed host plants in the United States that arise from increasing adoption of genetically modified crops and land-use change [meaning high herbicide use], not from climate change or degradation of forest habitats in Mexico. Therefore, reducing the negative effects of host plant loss on the breeding grounds is the top conservation priority to slow or halt future population declines of monarch butterflies in North America."

Exacerbating the problem of milkweed decline, the Energy Policy Act of 2005 and the Energy Independence and Security Act of 2007 give incentives, like subsidies and quotas, to farmers to convert their land designated as Conservation Reserve Program acreage into corn and soybean fields for biofuels—predominantly ethanol, which is made from corn. Grasslands in the Midwest are disappearing at the same rate as rainforests in Brazil.

One place where grassland habitat can be maintained is along the sides of more than four million miles of public roads in the United States. Roadsides have the potential to be important corridors of habitat for monarchs and pollinators that share their habitat. Citizen scientists in the MLMP have firsthand experience in finding monarch caterpillars on milkweed, often on public lands, including roadsides. When Patty Moss in Port Colborne, Ontario, couldn't find any monarch caterpillars on her property in 2013, she noticed a monarch laying eggs on milkweed along the road. As she recounted for the MLMP newsletter, "That butterfly started a mission of ours to protect monarchs alongside our rural roads." That year she took monarchs eggs from the roadside and raised and released over a hundred caterpillars. The following year she surveyed monarchs along nine road-sides. Because of imminent mowing, she rescued and rereleased over 250 caterpillars. Knowing a better solution was to delay roadside mowing, she advocated on behalf of monarchs to the Port Colborne City Council, which approved delaying the mowing from July 1 to October 1. Paul Lipman had similar success advocating for monarch-friendly mowing times with East Leverett Meadow in Massachusetts. Sometimes all it takes is an e-mail. Nancy Werner of St. Paul, Minnesota, e-mailed the Minnesota Department of Transportation about the importance of milkweed habitat along a frontage road where she monitored monarchs, and the department shifted mowing times to avoid the monarch breeding season. Monarch conservation is piecemeal, with individual citizen scientists taking initiative to make a difference where they live.

Other well-meaning monarch enthusiasts and gardeners, in trying to compensate for lost milkweed, may inadvertently create new problems for monarchs. Because of the clear consequences of the loss of milkweed in agricultural areas, people are planting milkweed on their own lands. Unfortunately, the most common milkweed available at stores is tropical milkweed, which has a longer flowering season than milkweed native to the United States. Native milkweed will begin to die in the early fall, but the nonnative tropical milkweed, when planted in the southern United States, won't die back at all. When monarchs encounter vibrantly blooming tropical milkweed, they might stop their migration and sometimes even begin breeding. In this way, tropical milkweed could contribute to a reduction in or loss of monarch migration. More worrisome is that tropical milkweed can act as a reservoir for a protozoan parasite that is deadly to monarchs. Called *Ophryocystis elektroscirrha* (OE), this single-celled organism lives all over the world and can only live and breed in one type of host animal: monarch butterflies.

An infected female monarch will pass the parasite to her offspring because her eggs will be covered with OE spores. When a caterpillar hatches, its first meal is its egg's shell. Although the spores are dormant, the caterpillar's digestive juices will break open the spores, releasing the OE protozoa, which sexually reproduce as the caterpillar grows. When caterpillars turn into pupae, the OE sexually reproduce again, and heavy infestations often cause wing deformities. About three days before the adult is ready to emerge from its chrysalis, the OE will enter their spore stage, and when a monarch emerges it can be covered in millions of tiny spores. The infection will die when the monarch does, but the spores can remain on milkweed plants. Because native milkweed dies back each year, the next season of milkweed will be free of spores, but with tropical milkweed, the load of spores accumulates year after year.

To understand the disease, its spread, and how to reduce its prevalence, Sonia Altizer, a professor at the University of Georgia, started a citizen science endeavor called the Monarch Health Project in which participants use a cellophane tape method to sample the abdomens of monarch butterflies for assessment of OE infections. Wearing surgical gloves and holding a monarch adult by its wings, participants wrap the abdomen in a standardized oval piece of ultraclear tape provided by the project, press it firmly, and then peel it off. Some of the monarch's scales come off, but the butterflies will recover. OE will be on the scales if the individual is infected. Participants send their samples to the Altizer lab, where researchers view the sample under a microscope to rapidly classify the severity of spore infection. With data from volunteer collectors, Altizer and her colleagues found that monarchs without migratory behavior are more likely to have OE infection. In North America, only 8 percent of the migratory Monarchs east of the Rockies have OE, but there is a 30 percent infection rate in monarchs west of the Rockies. There are small populations of nonmigratory monarchs in southern Florida and Texas, and these populations have infection rates of over 70 percent. Altizer's team recommends that those growing tropical milkweed cut it back in November and keep it cut back until February to prevent a buildup of infectious OE spores.

· · ·

"Populations of Invasive, Deadly Insect Pest Threatened by Environmental Change," read no headline ever. Why do monarch populations decline while mosquito populations increase? Why do troublesome species become more common, and species we value become rare? Why do the good die young?

When monarchs are insect ambassadors, we see them on labels for non-GM products or use them as "poster" insects leading the charge for pollinator conservation. When mosquitoes—one of the comparatively few insects that are nothing more than pests—serve as the ambassadors, they ruin potential diplomacy for the rest of the insect world. Pests give all insects a bad reputation. This guilt by association leads to pest management with broad brushstrokes of general pesticides that can unwittingly wipe out nontarget species. For example, native bees, honeybees, and other pollinators can be killed by many pesticides aimed at insects that eat crops.

Citizen science efforts vary in their design, and many insect projects are focused on detecting trouble from invasive pests. State agencies have issued calls to arms, either to detect something good and rare, or to stop an invasion of something bad. Mosquitos, for example.

There are thousands of species of mosquitos. Most do not have the ability to travel long distances on their fragile wings, and thus they are natural homebodies. Unfortunately, they can accidentally travel much farther in cars, in trains, and in trucks carrying food shipments. One hitchhiker on the most wanted list is a species with the scientific name *Aedes albopictus*—the Asian tiger mosquito. This mosquito is not just an itchy annoyance; it spreads diseases like West Nile virus and malaria, and is responsible for many cases of chikungunya in France and Italy and dengue fever in Croatia and France.

Since the 1980s, *albopictus* has spread across the Mediterranean region and into the northern part of the European Union. In 2011, while summer vacationers were traveling in Germany, some unwittingly carried *albopictus*. The Asian tiger mosquito was found in traps set by professionals, and monitored by volunteers, at several roadside rest stations. Germany immediately recruited volunteers for nationwide mosquito monitoring, and the citizen science project Mückenatlas (Mosquito Atlas) went online in 2012. Participants collect mosquitos around their homes. Squashed mosquitos are difficult to identify, so volunteers collect the pests intact and, undoubtedly with much satisfaction, chill them to death in the freezer next to the ice cream. The collectors then mail the thawing mosquito carcasses to designated researchers who can identify the species based on morphology and/or genetics. Researchers report back to participants and let them know which one of the potential thousands of mosquito species they found. Only by involving numerous volunteers can researchers hope to coordinate comprehensive surveillance and data collection useful to understanding and predicting where the most deadly mosquitos will be uncovered. The same protocols are used in Spain and will likely expand to all the European countries.

In the United States, invasive mosquitos also transmit canine heartworm, dengue fever, malaria, West Nile virus, and yellow fever, among other diseases. Lee Cohnstaedt, a research entomologist with the US Department of Agriculture (USDA), helped launch the Invasive Mosquito Project to focus surveillance on two invasive species of concern in the United States: *Aedes albopictus* (because it transmits dengue, chikungunya, and equine encephalitis) and *Aedes aegypti* (because it transmits yellow fever, dengue fever, and chikungunya). Some worry that *Ae. Aegypti* could spread the Zika virus across the eastern United States.

The USDA's strategy to reach national-scale surveillance is to recruit teachers to adopt the project into their curriculum. Kids throughout the country are learning to trap mosquito eggs and send them to the USDA. According to the protocol, each student lines two plastic cups with a labeled paper towel, fills the cups with water, weighs them down with a few rocks, and places one in a sunny place and the other in a shady place. After leaving the cups for seven days, they remove the paper towels (called germination paper) and let them dry, which prevents any mosquito eggs on the paper from hatching. Students identify, count, and record the different types of eggs laid by different species of mosquitos, and then package up the germination paper in a zipper-top food storage bag; they then send their data and the germination papers directly to the USDA. The data are helping the USDA keep tabs on what mosquito species are where and to track population fluctuations. The results will help improve maps used by the Centers for Disease Control and Prevention and will assist in mosquito management.

In the United States, the public also plays a role in the surveillance of diseases spread by other insects. In 1941, for example, Sherwin Wood, a professor at Los Angeles City College, recruited miners in Arizona to collect insects with the jingle, "Nab that bug at one cent each, for Wood at City College to keep." Today volunteers collect kissing bugs to assess the public health risk they pose. Kissing bugs, also known as cone-nose bugs or chinches, tend to bite us around the mouth while we sleep. About an inch long, they have a skinny seahorse face (which is mostly mouthparts) with long antennae, long skinny legs, and a flat, wide body that looks like someone already had the pleasure of squishing it. There are eleven species in the United States, most of them in the warmer climates. At Texas A&M University, Sarah and Gabriel Hammer run a lab that studies kissing bugs. Their graduate student, Rachel Curtis-Robles, takes kissing bugs that volunteers freeze and mail to her and screens them for *Typanosoma cruzi*, a parasite that causes Chagas disease in dogs and humans, as part of her dissertation research. Chagas is most prevalent along the US-Mexican border; most states don't track the disease, so statistics are hard to come by, but based on volunteer contributions, the Texas A&M group estimates several hundreds of thousands of cases in humans in the United States.

If British entomologist Helen Roy could choose an insect icon or ambassador, it would not be a monarch, mosquito, or kissing bug; her choice would be the ladybird (called the ladybug in the United States). Her favorite ladybird species is one that doesn't have a common name but goes by the Latin *Nephus quadrimaculatus*. Only twenty-six of the forty-six ladybirds that are native to the United Kingdom have common names; the named are easily recognized as bright and shiny polka-dotted ladybirds, while the unnamed are very tiny (one to three millimeters in size) and covered in hair that dulls the distinctive polka dot patterning. Roy speaks ardently about *Nephus quadrimaculatus*: "It is quite striking when you look at it . . . under a microscope."

Roy uses observations from volunteers, called biological recorders, from the UK Ladybird Survey (and other similar volunteer-led recording schemes and societies) in her work at the Centre for Ecology & Hydrology,[32] where her scientific research informs conservation and government policy. She also runs the Ladybird Survey, and that's in her free time, as a volunteer. As she puts it, "I'm a scientist and a citizen scientist."

The most invasive species of ladybird, the harlequin, has spread from Asia across Europe after introduction as a biological control agent for aphids, arriving in the United Kingdom during the summer of 2004. Based on observations reported by ordinary people spending time in their English gardens and the countryside, Roy mapped the harlequin's outward spread from the southeastern United Kingdom around London to the north and west.

As a result of human travel and trade, every continent has nonnative species, but a species is considered invasive if it has the potential to cause harm to ecosystems or human health. For example, in the United States, research suggests that there are about fifty thousand nonnative species in the country today, of which just over four thousand are considered invasive. Roy tells me that it is generally accepted that anywhere on the globe, only about 5 to 15 percent of the species introduced to an area have sufficient negative effects to warrant the label *invasive nonnative species*. When it comes to ladybirds, Roy re-

---

[32] In 1964 the Biological Records Centre (which is now part of the Centre for Ecology & Hydrology) was established in recognition of multiple recording schemes, all led by volunteers, on such topics as bees, wasps, ants, snails, and fish. There are now more than eighty such schemes.

assures people to not worry about the vast majority that have moved beyond their native range. For example, the bryony ladybird is non-native in the United Kingdom, but it feeds only on white bryony, "so no worries," she declares. The 15 percent or fewer causing trouble, however, have truly devastating negative effects. For example, the harlequin ladybird eats some other ladybird species, outcompetes the rest, and is resistant to diseases affecting the natives. The harlequin can carry, transmit, and not succumb to a fungi deadly to native species. Roy published a paper in 2012 that used citizen science data from across the British Isles and Belgium demonstrating that seven of the eight native species of ladybirds (those with good enough data for such an analysis) were declining because they are at risk from the spread of harlequins ladybirds.

One of the survey volunteers is Sheila Norris, who had been reporting ladybirds to Roy for about three years when, on vacation in Yorkshire, she found a harlequin ladybird. Her home is on the Isle of Man, and she and her Manx neighbors (as natives on the Isle of Man are known) started finding harlequins on the island shortly after other volunteers documented their appearance in Liverpool, which is about a three-hour ferry ride away.[33]

Norris is sixty-seven, and her typical day involves checking in on her ninety-two-year-old father, gardening, learning Manx Gaelic, and walking the countryside.[34] About three years ago, she noticed an odd ladybird. An Internet search led her to the UK Ladybird Survey, and she's been sending in her sightings ever since. "I think I was very encouraged by the fact that Helen Roy acknowledges every single record

---

[33] Norris is considered a "come-over" because she was born near Birmingham and "came over" to the island. She tells me that some people would say you need four full Manx grandparents to be a true Manx.

[34] Norris also volunteers for the Manx Wildlife Trust, and, like many citizen scientists, many other programs. She spent four years contributing to the Flora Survey (surveying all flowering plants in a two square kilometers), which will help create an atlas of flora for the Isle of Man. She contributes to butterfly surveys, and a few years ago she learned to identify fungi at a workshop held by a leading mycologist, so she adds that to her surveys, part of the Manx Biodiversity Partnership.

sent in personally (not just a computerised thank you)," she explained in an e-mail.[35] "I have to admit that it is a therapy for current family anxieties and responsibilities. It is also something entirely different from the rest of my career." Roy does enjoy every glimpse of ladybirds reported from around the country. She says she closes her eyes and conjures up the images in her mind to vicariously live the experience of discovering ladybirds.

The UK Ladybird Survey started in 1968, just before Roy was born. "I'm standing on the shoulders of those giants who began the project," she says. "I feel fortunate to have inherited the glory years, thanks to smartphones and the Internet."[36] There are around fifty thousand recorders involved in the survey, seventeen thousand of which are core participants whose records are easily verified either because they send photos or have a reputation as known, seasoned recorders. "I feel quite lucky to get sustained participation from so many recorders who are so different," Roy comments. "They're a diverse set, including schoolchildren, parents, gardeners, and amateur naturalists." Some work in local groups but, by and large, most don't know one another. But they all know Roy. Norris notes, "I've never met her, but after 3 to 4 years of participating in the scheme she almost feels like a friend! She makes recorders feel appreciated."

The common names of ladybirds tend to reflect the number of their spots. Norris's local reserve was a ladybird hot spot in 2014 with four abundant species: the two-spot, seven-spot, ten-spot, , and eleven-spot ladybirds. Nearby, she and a friend found fourteen-spot and eighteen-spot ladybirds as well as the orange and the larch ladybirds. As more people look for these insects, the records for each county improve. Participants are excited when they learn that their observation of a particular species was the first record of that species in their county. Often it's not because the species had just arrived in

---

[35] Roy receives about twenty-five thousand records per year. During the busy season, she might get fifty records a day. To respond to those with correct identification, she does use a standardized reply, which she tailors each month with a natural history theme.

[36] Roy accepts records via an iPhone app, online recording forms, and postal mail.

the county but because no one had yet taken the time to document its presence.

In North America, there are almost five hundred species of ladybugs, of which about seventy-five are the big and shiny polka-dotted ones. In the United States, the populations of three native species of ladybugs have crashed: the two-spotted, the nine-spotted, and the transverse ladybug.

Until the mid-1980s, the nine-spotted ladybug was the most common ladybug in the US Northeast, and it was declared the state insect of New York in 1989. By 1993 it could not be found in New York, nor in any of ten other nearby states. Now it is the target of a nationwide hunt called the Lost Ladybug Project, which began in 2000 but didn't molt into full instar until 2008, when Internet and mobile technology enabled photographs to start easily rolling in.

Replacing the nine-spotted is the invasive European seven-spotted lady beetle,[37] which is itself battling replacement by the harlequin ladybug as the latter makes its way across the United States, where it is known as the multicolor Asian ladybug. (It's called multicolor because there are easily thirty to forty different color variations, known as color morphs, in this species.[38]) Taking into account the contentiousness of terms regarding race, Leslie Allee, the outreach coordinator of the Lost Ladybug Project, prefers to call the multicolored Asian by its genus name, *Harmonia*—a distinct irony, because this invasive insect does not create harmony. Allee, wearing a white tailored shirt with a ladybug print when I met her, explained that there is a dearth of field observations of ladybug species. This limits scientists' ability

---

[37] The seven-spot (from Europe) has earned the title of state insect for Delaware, Massachusetts, New Hampshire, Ohio, and Tennessee.

[38] Based on volunteer data in the United Kingdom, Helen Roy examined whether the blacker morphs of the harlequin provided a thermal advantage in certain habitats. Looking at the spread of black and orange morphs across habitats, she found almost no difference in their spread, except the orange morph did particularly well in dark, coniferous woodlands.

to really know what is happening between the native nine-spotted ladybug and the two invasive species, the *Harmonia* and the European seven-spotted ladybug.

The first people to find a nine-spotted ladybug in the eastern United States were two young siblings in the greater Washington, DC, area in 2006. Jilene and her brother Jonathan Penhale (ages eleven and ten, respectively) found a single nine-spotted ladybug near their home. Subsequent searchers found no more, even after several events—called BioBlitzes—attracted hundreds of volunteers to scour the area and record every living thing they saw. Researchers have concluded that there is not a viable nine-spotted population in northern Virginia.

In 2011 a viable population of nine-spotted ladybugs was discovered on property of the Peconic Land Trust on Long Island. The site has been a community-supported agriculture (CSA) farm called Quail Hill Farm for decades. The farm hosted some ladybug hunts, during which Peter McDonald, a CSA member leading a group of kids, found nine-spotted ladybugs.

The Lost Ladybug Project has collected more than eleven thousand images of ladybugs, and over forty of those are of individual nine-spotted ladybugs. When a large population of nine-spotted ladybugs was discovered by a mother and daughter in Oregon, researchers noticed that they were smaller than historical specimens in museums. The bigger the ladybug, the more aphids it eats, and the better it functions as a control agent in integrated pest management. The undersized nine-spotted ladybugs are probably not as effective at controlling aphids on crops as are ones of normal size. Was the drop in size a genetic change, or a response to poor growing conditions locally? Was it caused by less food, new diseases, or predators preferentially eating the big ones? To find out, researchers collected individuals from the populations newly discovered by citizen scientists and started a captive breeding program with experimental studies.

I visited the captive breeding colonies of the Lost Ladybug Project at Cornell University and saw nine-spotted ladybugs from Quail Hill

Farm, others from Oregon, and a few from Colorado. I had imagined aviaries scaled down to not much bigger than an ant farm. Instead, the captive breeding facility looked like a restaurant kitchen. The colonies are kept on metal standalone shelves along the walls. On the shelves are cafeteria trays in varied shades of brown and gray. Each tray supports stacks of tiny plastic containers, the type that would be used for takeout condiments like mint chutney or salad dressing. The trays are sorted according to ladybug age groups. Each container holds one adult, a tiny, folded, damp piece of paper towel, and aphids. On another tray are stacks of containers that hold larvae, a tiny, folded, damp paper towel, and aphids. Some of the larvae are entering their pupae stage. On another tray are containers with only eggs and the tiny, damp, paper towel. To mate the ladybugs, they place a male and female together in a take-out condiment container for one day. On the table is an old mixing bowl teeming with florescent green aphids; a few spill over onto the table. The aphids are reared on fava bean plants in a small neighboring greenhouse room with a growth chamber. Each leaf of the fava bean plant supports millions of aphids. An undergraduate assistant sweeps the aphids gently into a mixing bowl. Then, using tweezers, pinches of aphids are dropped into each ladybug's container.

The stunted ladybugs collected in Oregon successfully bred in captivity, and researchers saw that the offspring grew to normal size (or even slightly larger), indicating that the small size of the rediscovered nine-spotted ladybugs was not a permanent genetic change. Further captive study found that lack of food affected both survival to adulthood and the size of adults. In the wild these insects were probably eating only five aphids a day, but normal growth would occur if they could devour up to twenty aphids per day. Where were the aphids? Most likely they had been eaten by seven-spotted ladybugs. *Harmonia* is also more effective at eating aphids than nine-spotted ladybugs, but it is unlikely to be suppressing the size of Oregon nine-spotted ladybugs through competition because *Harmonia* has not yet fully spread to the western United States.

Even if different species appear to do the same job, such as eating aphids, retaining biodiversity is necessary to ensure an ecosystem resilient to environmental changes. First, research has shown that pest control is better with a more diverse community of insects. Second, each species has its own niche in nature and fills a spot in the web of life of one species eating another, which means their loss has negative consequences that can ripple through these interconnections. Third, having a backup species, even though no two are entirely interchangeable, is better than nothing in case a disease wipes out the primary one. The bottom line is that biodiversity is a solid insurance plan for ecosystem health and services.

The same is true for citizen scientists: diversity is key. Citizen scientists include children, convicts, and senior citizens. James, the entomologist in Walla Walla, Washington, was heartened when inmates took deeper meaning from their butterfly experience and gained optimism about their own possible transformation. Butterfly metamorphosis is an easy metaphor; the inmates would comment, "We can change too." Given how well the inmates care for butterflies, perhaps they do change. Insects are not just valuable for the ecosystem services they provide but for their positive effect on humanity's psyche. Yet insects lead the ranks in what appears to be the sixth major extinction event in the earth's four-and-a-half-billion-year history. If butterflies were to look at humans as metaphors, would they see our hands in their doomed future or in an impending metamorphosis?

Citizen science is often based on a division of labor: the field or lab work is fun, rewarding, and/or relaxing enough to warrant dedicating leisure time, while nonleisurely activities, like statistical analyses, are carried out by paid scientists. Yet for a less common sort of citizen scientist—called the amateur scientist—everything a professional does is compelling enough to warrant pursuit. Inmates can be not only citizen scientists but also amateur scientists. Robert Stroud, for instance, became known as the Birdman of Alcatraz. While serving part of his life sentence at Leavenworth, Stroud raised and studied over three hundred canaries and published two books about avian diseases.

Some amateur scientists study birds, and some study butterflies. Ann and Scott Swendel, a couple who have dedicated their free time for the last thirty years to the pleasure of carrying out scientific research, study both. They do not have the credentials of scientists, academic affiliations, or funding for research. Nevertheless, they have authored dozens of peer-reviewed publications in respected entomology and ornithology journals. Ann Swendel explains, "Science is determined by the nature of the activity or product, not the person who did it."

In astronomy, amateurs have had a long and profound history of investigating space and the worlds around our world. Given that modern astronomy involves expensive technology, from big ground-based telescopes on our highest mountains to orbiting satellite telescopes, what can people do with backyard telescopes, dark nights, and good luck that professionals can't? Chapter 4 takes up that question.

CHAPTER 4

Astronomy
The Pull of the Planets

If you know what you're looking for, that's all you'll get—
what's previously known. But when you're open to what's possible, you get something new—that's creativity.

—ALAN ALDA

I N OUR UNIVERSE, IT IS THE ENTERTAINERS AND ATHLETES WHO RISE TO
fame. Imagine a parallel universe where scientists receive fanfare.
Stretch your imagination further. In which galaxy might the icons of
the silver screen seek out the autographs of citizen scientists?

Our own Milky Way.

Such were the events of 2005 for New Zealander Jennie McCormick.
While at work, McCormick overheard someone inquiring for her. She
turned around and was approached by a tall woman with strawberry
blond hair, a star of the Hollywood variety. Because it was unexpected,
and because the enormity of the cosmos has a way of making people
feel small and insignificant, the next few minutes are a blur in
McCormick's memory. Gates McFadden, the actress who played Com-
mander Beverly Crusher, MD, in *Star Trek: The Next Generation*, was
in New Zealand for a science fiction convention and wanted to meet
McCormick. McFadden had heard about McCormick's discovery of a
large planet thousands of light years from Earth.

When McCormick's heart rate lowered and her memory came back,
she was holding The New Zealand Astronomical Yearbook 2005 and
it was signed, "To Gennie [*sic*], from the doctor in space to the *real* discoverer who went *boldly* where no one went before." McCormick gave
McFadden a book, and wishes she could remember which one and how

she signed it, but details of that day escape her. The large planet was her first big find, but McCormick went on to see meteors, fireballs, comets, and even a solar system that bears remarkable resemblance to our own. Yet, she confides, "Gates McFadden wanting to meet *me*—that was the most amazing thing of all."

McCormick is not a professional astronomer, nor a science celebrity like Neil deGrasse Tyson or Bill Nye the Science Guy. She is an amateur astronomer. A stellar amateur, one might say. She's been one for twenty years, and has an observatory at her home in Farmcove, a suburb of Auckland. The term *amateur* is a longstanding, nonderogatory phrase in astronomy. If she was an amateur actress, amateur poet, or amateur singer, those would be fighting words, implying that she was a hack, but to be an amateur in astronomy is to have enormous skill, dedication, and expertise akin to that of amateur golfers, Olympic athletes in the 1900s, or amateur cricketers in the United Kingdom.

Nevertheless, one amateur society dating back to 1911, the American Association of Variable Star Observers, prefers the term *citizen astronomer*. McCormick says she isn't sensitive about words, but her hackles are raised a bit by the term *amateur* when we speak via Skype. At the Farm Cove Observatory website I find a photo of McCormick; there she is, smiling in a baseball cap, jeans, and black T-shirt with pink lettering stating "It ain't easy being a bitch." Early in our conversation, she urges me to call Andrew Gould, a professor at Ohio State University who uses McCormick's observations, and presses me to ask him certain questions: "What is professional astronomy? Professionals and amateurs do same thing. Is a professional someone who is paid for it? Someone with a PhD? Ask Andy what he thinks a professional astronomer is!"

The day before, I had already asked Gould about amateurs. Specifically, I'd asked about the biggest discovery in astronomy that was made by an amateur. His answer was Clyde Tombaugh's discovery of Pluto as a planet in 1930. But look how that turned out! Pluto held the status of planet for about fifty years and became the namesake of a beloved animated Disney character before being demoted to a minor planet.[39] Most articles about Tombaugh refer to him as a professional astronomer, but if Gould's opinion is typical, some astronomers in academia consider Tombaugh to have been an amateur astronomer. Tombaugh was employed by Lowell Observatory as an observer; he earned his undergraduate degree and master's degree in astronomy, and even taught astronomy, but he did not earn a PhD. Tombaugh did what amateurs do, spending lots of time observing so they can find asteroids, supernovas and, in this case, a new planet that would later be demoted to minor planet status.

If the fellow who discovered Pluto is considered an amateur, then all that's left to distinguish an amateur astronomer from a professional astronomer are credentials and academic employment. Like the Birdman of Alcatraz, amateurs are unique among citizen scientists in that they carry out independent research, just like professionals do. In addition to their independent research, they often participate in research networks with professionals and other amateurs.

In astronomy the amateurs generally own telescopes fitted with equipment similar to digital cameras for recording observations. McCormick was shepherded into amateur astronomy by the Center for Backyard Astrophysics (CBA). All CBA amateurs must have a telescope with an eight- to twenty-six-inch mirror and a motor so it can track objects as the earth rotates, and a charge-coupled device (CCD) mounted on the telescope, which is the heart and soul of a digital camera. Of course they need a computer, e-mail, and software to take hundreds of images and string them together into a time series. Amateurs have to know how to read star charts and online resources like star surveys. Computers and the Internet are important, but of all these technologies, the digital camera is what revolutionized astronomy— both professional and amateur. Instead of using old-fashioned cameras with films of photosensitive chemicals that darken when exposed to

---

[39] Beginning in the early 1990s, astronomers began discovering many similar objects near Pluto, in a region called the Kuiper Belt. In 2005, Eris was found, and it was bigger than Pluto. To make sense of the distant objects orbiting the sun, astronomers created a formal definition of planet, which excluded Eris and Pluto.

light, the CCD enables the passive collection of photons that are instantly stored as electrons and shifted into place to make an electronic picture. The technical term is photometry (collecting data from photons, or starlight).

The financial investment, sleepless nights, and mental effort seem greater for stargazing than for birdwatching, butterfly tagging, or weather recording. The higher bar for participation lends itself to a different experience. While many of us enjoy nature nestled under a warm blanket of blue sky, amateur astronomers wait until the blue sky fades, like a curtain being drawn open to reveal the enormity of deep cold space that surrounds us.

What about millions of others whose curiosity extends beyond earth into the depths of space, but who don't own telescopes? Such citizen scientists don't necessarily wait until dark. With the Internet, they explore spaces in their free time, day or night, typically by reviewing and classifying archived data of outer space, labeling images, spotting signatures in light curves, and tracing features on photographs such as craters on the moon.

The most frequently explored astronomy data on the Internet is from the Sloan Digital Sky Survey (SDSS). This Sloan Foundation survey began with a 2.5-meter telescope at Apache Point Observatory in southeastern New Mexico in 2000;[40] it has been continually gathering data ever since. These data produce the most highly detailed three-dimensional map of the universe; the key to their value is also in their expert storage and management, which makes the data easily accessible. In the decades before the Internet, astronomers stored observations on tapes; to access the data one had to travel to the physical archive and retrieve the tape. Alex Szalay, an astronomer at Johns Hopkins University, taught himself computer science and collaborated with Microsoft specialists to design an online digital archive for the SDSS that made the data easier for anyone to access than any astronomical data at the time. Instead of being stored on tapes, the

---

[40] The lens of the telescope is 2.5 meters in diameter.

data were on a computer server accessible via the Internet. Fortunately, as we'll soon see, the data were also made redundant on backup servers.

An unusual illustration of excellent project design and management, the SDSS collected data of a higher quality than necessary for its simple mapping mission. For one-third of the sky, the images are multicolor, and for more than three million objects there are graphs of the brightness of the object at different light wavelengths; these graphs are called spectra. Because the images and spectra are easily available to astronomers via servers, and many of the objects are of particular interest, the SDSS is to thank for over 5,800 scientific publications,[41] each an incremental advance in understanding galaxies, quasars, intergalactic gas, and more. About fifty of these publications were possible because of an online citizen science project called Galaxy Zoo that launched in 2007.

At the launch of Galaxy Zoo, the SDSS had about a million images in the main sample, and computer algorithms were not handling the nuances of classifying galaxies sufficiently. In particular, the algorithms were not detecting elliptical blue galaxies, which Kevin Schawinski, a PhD student at the time, was extracting from the SDSS data. Most elliptical galaxies are red because they happen to be dominated by older stars, while most spiral galaxies are blue because they happen to contain young, actively forming stars. But redness and blueness do not reliably distinguish elliptical and spiral galaxies. Elliptical blue galaxies are uncommon, and Schawinski believed they would provide a key missing link for understanding the evolution of galaxies. He started classifying SDSS images all day, every day. By the end of the first week he was bleary-eyed, but had classified fifty thousand images: one classification every four seconds for eight hours every day for seven days. He found a few hundred elliptical blues and started a database called Morphologically Selected Ellipticals in SDSS, or MOSES (astronomers have a penchant for acronyms; NASA even has acronyms for

---

[41] The SDSS has been more scientifically productive than even the Hubble Space Telescope.

acronyms).[42] With this sample, Schawinski got an inkling that there were sufficient numbers of elliptical blues to study. With a sample size of billions and billions of objects, even rare objects could appear in large numbers. Graduate students are known for taking on insurmountable tasks, but Schawinski was determined to find another way to sift through a million images.

At the point when he was stumped and ready to duct tape undergraduate students into chairs until they completed their fair share of classifications, he happened to meet Chris Lintott, a well-respected and charismatic astronomer known for effective public outreach. Over beers at the Royal Oak Pub in Oxford, England, they reasoned that the public was already strapping themselves into chairs at computers, so why not put the images online and ask the public for help with the classifications?

They asked online volunteers to inspect galaxies and split them into ellipticals and spirals (the two primary shapes) and mergers (combinations of the two shapes). When volunteers encountered spirals, they were also asked to record the direction of the arms (clockwise, counterclockwise, edge-on/don't know). The rotation of the spirals was of interest to another Oxford University researcher, Kate Land. She had been planning to set up a laptop computer in the cafeteria of the Oxford Department of Physics to enlist others in sifting through galaxies and identifying the rotational direction of the spiral arms while digesting their steak and kidney pies. She came on board with the idea of asking the public instead.

At the time, NASA was running an online citizen science project called Stardust@home. Participants like Bruce Hudson, who was nearly bored senseless by daytime TV while recovering from a stroke, were spending hours upon hours volunteering online. About twenty thousand volunteers, most less motivated than Hudson, were identifying tracks left by interstellar dust in the samples that NASA gathered by passing small aluminum blocks coated with a gel though the head of

---

[42] For example, NASA uses TLA to stand for three-letter acronym, and FLA for four-letter acronym.

comet Wild 2 (pronounced Vilt 2).[43] Lintott and Schawinski made the optimistic estimate that Galaxy Zoo might attract a quarter of the number of participants as Stardust@home. If each volunteer could classify just one galaxy per day, they figured it would take about three years to crowdsource classifications on a million galaxies in the SDSS sample. They wanted each galaxy visually inspected and classified by five different people, and would trust only those classifications that reached consensus.

When I spoke to Schawinski via Skype, he reflected that they were lucky with the initial web design choices, which were clean and simple. They also had luck in the timing of the launch of Galaxy Zoo, which was during an ebb in current events. BBC News picked up their press release, and the announcement of Galaxy Zoo's launch became the second most popular story shared that day.[44] The public response was orders of magnitude greater than they expected: to their astonishment, within twelve hours of the launch they were receiving twenty thousand classifications *per hour*, cumulatively surpassing the accomplishments of Schawinski's painful week of full-time classification. After forty hours they were receiving sixty thousand classifications per hour. In the first fifty days, there were over twenty-five million classifications. In the first year, over 100,000 people classified about 900,000 images, with each image receiving an average of thirty-eight classifications (and all with at least twenty). Schawinski was able to complete his dissertation in a reasonable length of time and move on to a faculty position at the University of Zurich's Institute of Astronomy.

The swell of people logging on to Galaxy Zoo in the first three hours overloaded the capacity of the Internet cable for the Sloan Foundation's servers, which were housed at the Johns Hopkins University's Fermilab. Schawinski says that Galaxy Zoo would have died right then if Szalay had not already cloned the server. Now the data are stored on cloud servers to avoid similar problems, and they still get bumps in use after

---

[43] This was the first sampling of solid extraterrestrial material beyond the moon.
[44] The Galaxy Zoo launch ran a close second to the story of a man who flew from Toronto to Wales for a friend's wedding, only to find that he was a year early.

media coverage of new findings from the project. Lintott and Schawinski instantly realized not only the enormous potential of crowdsourcing for help but also the public's intense interest in contributing to science (or, perhaps the publics' equally intense boredom with daytime TV).

Zooites, as volunteers named themselves, classified galaxies as reliably as Schawinski or any other professional. Later versions of Galaxy Zoo have required more choices of galaxy morphology and more complexity, but the volunteers have always stepped up to the tasks. It's not that every person got every classification right; there are systematic biases in the data. For example, Land tested for bias in identifying the rotation of spiral arms by having some Zooites classify mirror images of galaxies. She found that people were more likely to identify spiral arms as counterclockwise rather than clockwise. She speculated that the visual bias could be related to the predominance of right-handedness in the general population. At any rate, the bias could be addressed and the data used. Land was able to confirm that galaxies have no preference in the direction of their spiral and occur with equal likelihood as clockwise and counterclockwise.

Over time, the consensus tool has been refined. The number of classifications needed to reach consensus varies from project to project, and even within projects, but in the end the crowdsourced data sets are reliable and enable discoveries that would otherwise be impossible.

When I ask Schawinski to pick the most important finding from Galaxy Zoo, he splits the answer in two. First, the crowdsourced observations have been used by researchers, like him and his students, to identify transition stages in the evolution of galaxies. Those stages can now be contextualized, advancing our understanding of how galaxies evolve. About a thousand elliptical blue galaxies were classified by Zooites, and the subsequent discoveries about the evolution of galaxies laid the foundation for Schawinski's continuing research.

Second, the Zooites discovered a different set of findings of rare classes of objects. Schawinski explains that Zooites were finding things that no one even thought to ask them to look for. He quotes former

US secretary of defense Donald Rumsfeld, who once said that "there are known knowns; there are things we know we know. We also know there are known unknowns; that is to say we know there are some things we do not know. But there are also unknown unknowns—the ones we don't know we don't know." Zooites had found unknown unknowns.

The poster child for finding unknown unknowns in Galaxy Zoo is Hanny van Arkel, a schoolteacher in the Netherlands. While classifying galaxies, she asked, "What's the blue stuff below the galaxy?" This rare type of object became known as Hanny's Voorwerp (Hanny's object). Further investigations revealed it to be a quasar ionization echo near the constellation Leo Minor.

There is a structural feature of Galaxy Zoo that supports serendipitous findings, and this feature was itself a stroke of more good luck. A deluge of e-mail inquiries rivaled the data deluge that Galaxy Zoo was established to address. The larger-than-expected number of volunteers exceeded the capacity of the Galaxy Zoo team to respond to each. As a quick-fix solution, they established an online forum to send answers to every question to everyone at once and for participants to help each other with the assigned tasks. No one expected participants to go off script and make their own discoveries. Schawinski, Lintott, and the Galaxy Zoo team were happy to learn that once you get people talking, they start collaborating, and they undertake the scientific method to figure things out together.

Miranda Straub, a PhD student in science education at the University of Minnesota, retrospectively examined the forum archives within Galaxy Zoo to chronicle the serendipitous discoveries. In the forum, Zooites quickly noted which objects they found interesting and, as online communities do, they formed informal groups and decided on a plan of action based on their shared interests. When they needed input from scientists they asked for it, but otherwise they put their heads together. The majority in Galaxy Zoo simply click-click-clicked through the standard procedures. The more ambitiously curious used the forum as a place to discuss objects of mutual interest. They created threads dedicated to collecting objects with similar characteristics, the way kids sort the masses of Halloween candy before trading. One such group of objects were small, round, green galaxies. The first observations of these galaxies began as a lighthearted pun on the forum: "Give peas a chance." Others responded with gags about making soup. As people processed images, if they noticed a galaxy that looked like a green pea, they linked it to the appropriate forum thread. Schawinski noticed their interest and explained to the Zooites that they needed to look at the spectrum (the brightness at each wavelength) of each object because spectra revealed what made these "peas" unique. He communicated with them as he would colleagues, suggesting scientific papers to read. The information was complex: the mark of a green pea was doubly ionized oxygen emissions that produced the green color. Later, when a new Zooite using the online name starry nite asked on the forum how to know if an image qualified as a green pea, another volunteer, Rick Nowell, was able to explain what he'd learned from Schawinski and the papers and blog posts he'd read about the doubly ionized oxygen. Zooites interested in the green peas then started narrowing criteria visually and by spectra.

On February 2, 2008, starry nite found a true pea, and the post was full of excitement "A pea! A pea! Green, galaxy ID, correct spectral chart, not posted before! PEA! PEA! PEA! PEA! PEA!"

Not too long after, a Zooite known as FermatsBrother wrote code to automatically extract images likely to have green peas to narrow the search, leveraging the power of both humans and computers. Rick Nowell created a summary of possible green pea galaxies, and other Zooites on the thread added to the list of pea candidates. After they had been working to identify and find green peas for about a year, the Zooites had over a hundred possibilities and a core of thirteen highly likely peas. At that point, Carrie Cardamone, a graduate student, was assigned to moderate the forum. She started investigating the green peas found by Zooites, a number that soon grew to over 250. Cardamone worked with the volunteers and published the finding that these were rare dwarf galaxies with exceedingly high rates of star formation.

Around the same time that Galaxy Zoo was forming, two astronomers in Illinois were coincidentally venturing into citizen science on individual pathways that would eventually intersect. Pamela Gay, at Southern Illinois University, was heavily involved with the American Association of Variable Star Observers, a network of professional and amateur astronomers who coordinate their observations of the changing brightness of stars. After Gay met Lintott at the Astronomical Society conference in 2008, they began writing grants to join forces in the creation of an online site for more citizen science projects. Meanwhile, Lucy Fortson, then vice president for research at the Adler Planetarium in Chicago, was helping high school students across the United States use data from the SDSS to first identify quasars and then request follow-up images of these objects from robotic telescopes in Australia and southern Wisconsin. The board of the Alder was questioning the potential of Fortson's research carried out by high schoolers. They wondered whether the planetarium should dedicate an entire division to research, particularly of this sort, when the University of Chicago and Northwestern were so close by and had outstanding researchers. What, they asked Fortson, could Adler do to put themselves on the map next to these astronomy powerhouses?

"My answer," recalls Fortson, "was that I had a different model of carrying out research: the universities had the traditional professor, graduate student, and undergraduate model. The Adler had the general public, high school students, and teachers."

Looking at the teams objectively, most would probably play it safe and place their bets on the professionals. Even though the Adler board was skeptical that working with nonprofessionals could lead to publishable research, they accepted Fortson's proposal to begin a citizen science center at the planetarium.

At the time, Fortson did not know Lintott, Schawinski, or Gay, but the education liaison for the SDSS, Jordan Raddick, knew them all. Raddick connected them because of their common interests in public engagement in science, sparking the big bang that created Zooniverse, a hub of online citizen science projects, similar to Galaxy Zoo, that supports research in astronomy, other sciences, and the humanities.

Their shared idea for Zooniverse was to centralize online citizen science into one web-based portal in order to deal with the overflow of data in many disciplines. The Zooniverse cyberinfrastructure is also useful for crowdsourcing help after natural disasters. No one expects the floods of data to subside. Instead of trying to have citizen scientists replace computer algorithms, Zooinverse supports a system of research in which citizen science efforts are used to train machine algorithms; they combine human and computer computation. Gay was the first to get a US grant to fund several NASA-based projects that were hosted on the Zooniverse homepage. Now she runs a suite of NASA-based projects via an online portal called CosmoQuest.

Online crowdsourcing in astronomy arose because the infinity of space has led to a near infinite amount of data, and human minds have proven to be better at processing and sorting than computer algorithms alone. Plus, when amateur scientists handle the data, their labors not only help researchers use the data more quickly but the amateurs themselves make serendipitous discoveries. The same cannot be said about algorithms.

The telescope surveys feeding the data deluge do have a limitation: they capture only snapshots of space. Amateurs are vital to observing objects in space that move and change. Let's return to Jennie McCormick to learn more about how amateurs contribute to discoveries on their own as well as through networks of professionals and other amateurs. McCormick doesn't mine the data in the SDSS, but she does use the SDSS map to correctly position her telescope in order to make new observations.

McCormick dropped out of school at age fifteen, raring to be a jockey. She describes herself as a silly girl in gum boots, smelling of horse urine, with another ambition to be a veterinarian. Or, some days, a volcanologist. McCormick's interest in the night sky began when she was a young girl in Wanganui, a small town on the west coast of the North Island of New Zealand. The constellation Orion is, in

Greek mythology, a hunter with a belt and sword, but to Australians and New Zealanders his upside-down accoutrements look like a saucepan. "There's the pot," McCormick's mother would say when pointing it out. On her own, McCormick would steal looks at the stars while mucking out stalls at daybreak. While some kids may think, "I want to grow up to be an astronomer," the thought never occurred to McCormick, and she certainly never thought, "I want to grow up to be an *amateur* astronomer."

A broken jaw and several bad falls persuaded McCormick to leave horse racing for the slower pace of raising a family, all the while working to make ends meet. Nine-to-five jobs worked to pay the bills. During the other nine-to-five (from 9:00 p.m. to 5:00 a.m.), she began to observe the night sky. To date, she has published over seventy scientific papers in astronomy journals. At age fifty-one, McCormick has accomplished more than many scientists do in their lifetimes.[45]

In what ways do amateurs like McCormick contribute to astronomy? They help in areas of astronomy related to fleeting, often unpredictable, events, which includes detecting planets, variable stars, eclipses, asteroids, comets, and supernovas. Professionals compete intensely for the use of large telescopes; the selection process takes months, and at the end of it, each professional gets just a few sleepless nights with the device. With limited time, most can't spend night after night observing a transient situation, nor suddenly turn their attention to a particular region, the way amateurs can (albeit with less powerful telescopes). With observations from a few nights, professionals then spend years diving into analyses of the data. Gould tells me that most amateur data are of good quality—sometimes better than professional data. McCormick says amateurs produce high-quality data because they use their equipment all the time. In many cases, professionals' hands don't physically get on the big telescopes; the trend is for professionals to put their observations into a queue, where telescope staff carry out all the operations of the telescope. Professional astronomers

---

[45] What's more, she also published an ornithology paper based on her observations of tool use by the (very intelligent) magpies that hang around her observatory.

have to trust the data collectors, whether staff at a large telescope or amateurs with their own equipment.

Amateurs in possession of their own telescopes are also able to make better observations because they can make *more* observations: they are only limited by the whim of the weather, deterred by clouds. There are several ways the nonstationary and variable characteristics of the stars allow amateurs to make contributions that fall through the cracks of professional approaches to astronomy.

At a glance, space seems not only infinite but static. Toss away this notion. The Sun, the North Star, and all other little stars that twinkle twinkle and guide bird migration and sailor navigation are not actually stationary while we wonder what they are. Only within the context of our solar system is the Sun stationary, as Nicolaus Copernicus explained in *De revolutionibus orbium coelestium* (On the revolutions of the celestial spheres)—in 1543 (though he had shared the idea informally with close colleagues since 1514).[46] Most people accept that the Moon orbits the Earth, which orbits the Sun. Many don't realize that the sequence keeps going. The Sun orbits the center of our galaxy.

Our solar neighborhood is in motion, if slowly.[47] It takes our Sun hundreds of millions of years to glide around the center of our galaxy. The Milky Way Galaxy has the shape of a spiral with whirling arms, the thickest cluster of stars, gas, and dust forming a central protuberance, known as the Galactic Bulge. All the stars around us are orbiting the center of the galaxy in different paths. Some are inclined toward the center of the galaxy, some are elliptical, and some are just passing through while orbiting a different galaxy. Millions of years from now

---

[46] His was an idea light years (well, okay, about a century and a half) ahead of its time. Heliocentrism eventually replaced geocentrism, the idea that the earth is the center of our solar system.

[47] Astrometry is measuring positions of stars in order for them to be used as a reference frame. No single star can be a reference frame because everything is in motion. With a database of millions of stars, astrometry involves solving for the relative positions of many stars and their motions simultaneously—that is, a reference frame—taking into account movement of every single star possible.

earth will be in a different position relative to all other stars, and our descendants will see entirely different constellation patterns of stars in their night sky.

One enduring question that whirls in the minds of stargazers is whether there is life anywhere other than on Earth. Because it's dense with objects, the best place to search for extraterrestrial life is in the Galactic Bulge. The success of the search depends on the intersection of where planets are most likely to form, where planets are most likely to have life, and where we can best detect them.

Different types of planets form at different distances from their parent star. Gas and dust molecules condense to form dust grains that coagulate into planets beyond what astronomers call the snow line, the cold region far from a parent star. Like snowballs growing as they roll across the snow, a protoplanetary disk will grow bigger as gas diffuses across the snow line, the dust acting like sticky ice, building up more and more grains. The distance from a parent star may also influence the likelihood of the planet supporting life. Astronomers think that the planets that are most likely to support life are those with liquid water. Planets in the habitable zone are not too hot (or the water is always vapor) or too cold (or the water is always ice), but just right. There are three methods used to find planets in this Goldilocks zone.

One of these methods does not involve citizen science. The radial velocity method detects planets based on the Doppler effect produced by objects moving closer and then farther away. Stars with planets have a tiny wobble, as though the planet is speeding around the star like a race car approaching with a *vvrrrrr*—and departing with—*rrooommmmm*. In the 1980s, astronomers were measuring velocity in kilometers per second. The radial velocity method became popular during the 1990s when astronomers developed the ability to measure velocity with accuracy down to tens of meters per second.

The second of the three methods to find planets involves tracking brief dips in the brightness of a star. If a dip in brightness repeats at least three times at regular intervals, this indicates that a planet is passing between the earth and the star during each of its orbits. Ground-based telescopes could measure large dips in brightness (which is what happens while a small dull planet passes in front of a big bright star) and observatories in space, like the Kepler Spacecraft, could detect orders of magnitude smaller dips of planets farther away. One of the online citizen science projects in Zooniverse, called Planet Hunters, involves volunteers looking at graphs of the brightness of particular stars over time as recorded by the Kepler Spacecraft. Volunteers mark every dip in brightness, hoping to find evenly spaced repeated dips as signatures of an extrasolar planet. They can look for dips without an in-depth understanding of the details of how the data were gathered, processed, and displayed.

Planet Hunters do not typically have their own telescopes, nor the experience and skills to earn the title of amateur astronomers. There is a trade-off similar to balancing quality and quantity. There's a small quantity of high-performing amateur astronomers across the globe (perhaps a few dozen); there are several thousand volunteer Planet Hunters, and millions of citizen scientists across Zooniverse projects.

The third method is microlensing. Of the three methods, microlensing, which McCormick does, is the one most likely to find planets that other methods miss—namely, those near or outside the snow line (which tend to be big planets). Professional and amateur astronomers gather observations based on the celestial phenomenon of gravitational lensing as described by Albert Einstein's general theory of relativity.

Relativity reveals that the law of universal gravitation conceived by Isaac Newton in 1666 is not entirely universal. During the last epidemic of the bubonic plague, Newton left Cambridge to spend time around fewer people in his hometown of Lincolnshire. While viewing his mother's garden, as the story goes,[48] he saw an apple fall from its tree. Why did it always fall down? Why not sideways, or up? He concluded that the earth must pull objects toward it, a force he called gravity.

---

[48] I say "as the story goes" because some claim it stands to reason that Newton's account is actually a fable. Scientific discovery takes hard work. Scientists don't just sit in the garden and have ideas fall from the sky and bang them on the head.

Newton, who historians say was obsessed with the moon, also concluded that the earth must pull the moon toward us. Gravity, Newton codified into his law, must work over large distances.

But part of Newton's law does not hold up in the court of outer space: the principle that gravity exists only between two objects with mass. This line of reasoning passed muster for a few centuries. If one object has a mass of zero, Newton's law concludes that there is no gravitational force. Yet gravity exerts a force on light, which has no mass. Einstein's theory of relativity accounts for this phenomenon by introducing of the concept of space-time. Space is three-dimensional, and time runs in one linear dimension. If we rack our brains, we can consider space and time simultaneously, where time becomes a fourth dimension. In this scenario, every event has a unique position and unique time.

I'd speculate that Walt Whitman was contemplating space-time, not the unity of our shared human experience, in his 1855 poem, "Crossing Brooklyn Ferry": "It avails not, time nor place—distance avails not." Whitman stands in one spot and looks forward to people in the future and expects us to stand in the same spot to look back at him, face-to-face, if we could remove time. "What is it then between us?" Whitman asks. "What is the count of the scores or hundreds of years between us?" Einstein knew the answer: space-time.

The term *space-time continuum* has infiltrated science fiction, usually in reference to a threat, or maybe a thoughtless error, creating an imminent rip in the fabric of the entire universe. According to Einstein, the gravitational pull of objects with mass, like a big star or even a planet, bends the space-time continuum.

According to Newton's law of gravitation, the moon and earth pull each other, causing the smaller moon to orbit the larger earth. According to Einstein's theory of relativity, the moon is moving in a straight line through space-time while the earth's gravity bends space-time. It is as though the moon were a toy race car on a Hot Wheels drag strip, and the earth's mass caused the tracks to curve, and so the race car rounds the bend with the moon following that course.

The same applies to light rays. Light follows space-time. If a high-mass object causes space-time to bend, then light rays bend too. Light rays bending from gravity are no different from light rays bending as they pass through optical lenses. Thus, when a massive object bends space-time, it functions identically as a lens. Closer objects, which astronomers refer to as foreground objects, pass in front of background objects relative to our vantage point. A star moving in the foreground can bend light like a lens, giving observers a sharper view of a star in the background, causing it to act like a spotlight on any planets orbiting the foreground star. The transient placement of stars and their do-si-do spiraling around the center of our galaxy occasionally puts us in the right spot for lensing episodes, opening short windows of opportunity for detecting planets. Some lensing events are of low magnification and only detected by professional telescopes; others are of high magnification and detectable with telescopes typical of amateurs. For those of us on the ground, Einstein's theory means that, every so often, when certain celestial bodies—planet, foreground star, and background star—align just so with respect to our viewing, their mass can cause space-time to bend in a way that reveals to us planets and galaxies that are otherwise blocked from view.[49]

These celestial alignments, not predicted in the astrology section of any newspaper, only last about two weeks, a month at most. To detect planets requires continuous observations for as long a period as possible, which means astronomers on all parts of the Earth need to pitch in to assemble the full picture. The lensing events aren't very common, nor entirely predictable, so when they occur, Gould sends an alert to the Microlensing Follow-Up Network, called MicroFUN (or, with the

---

[49] There are also larger gravitational lensing events, which describe a background galaxy that becomes visible, smeared into a ring around a foreground galaxy. The ring is called an Einstein ring, a gravitational well around a spinning object. More typically, the entire ring is not visible; usually it is just four points, called an Einstein cross. Each point is multiple images of the same galaxy, like a kaleidoscope lens. Zooniverse hosts a project called Space Warps where participants identify Einstein rings in images.

~~Geek~~ Greek symbol, μFUN), which signals professionals and amateurs to gather as many observations as possible.

In MicroFUN, McCormick was one of two amateurs, along with thirty-one professionals, involved in detecting a planet orbiting a star in the constellation Scorpius. The planet was given the reference identifier OGLE-2005-BLG-071. In this identification system, OGLE stands for Optical Gravitational Lensing Experiment, followed by the year of the discovery (2005), and BLG stands for the Galactic Bulge (the center of the Milky Way Galaxy), followed by the numbered event of that year. It was the third planet discovered through microlensing. McCormick spent twelve hours collecting observation of the high-magnification lensing event for MicroFUN. The planet that MicroFUN was able to identify is three times the mass of Jupiter, about 540 million kilometers from its parent star, which is less than half the mass of the sun, and called a Red Dwarf. This planet's orbit takes about ten earth years.

While the enormous planet was a valuable find, McCormick's most important contribution to astronomy, and the most important discovery of MicroFUN to date, was reported the following year as OGLE-2006-BLG-109L and described in the prestigious journal *Science*. The discovery occurred over a two-day period. McCormick was unable to view the microlensing event for the first twenty-four hours due to clouds obscuring her view. Another amateur, Grant Christie, who lived fifteen kilometers down the road from her and whom McCormick had recruited into the network—had clear skies. McCormick was brewing in frustration at the meteorological injustice! Fortunately, the sky cleared in time for her to observe the tail end of the event. Unexpectedly, OGLE-2006-BLG-109L was not one planet, but two: a small-scale solar system with planets similar to Jupiter and Saturn in their relative sizes and distances from their parent star, located in the constellation Scorpius. McCormick had helped uncover what science fiction fans would call a parallel universe.

According to Gould, OGLE-2006-BLG-109L remains the only planetary system discovered to date that looks similar to our solar system. Our solar system appears three times richer in planets than other stars along the line of sight toward the Galactic Bulge. Catching two planets around one star was possible because a network of professionals and amateurs across the globe are on alert to observe as soon as someone in the network notices a lensing event. Because analog systems are so difficult to detect, astronomers can only roughly estimate how many exist. Gould estimates that one in six stars have solar-like systems with planets like Uranus and Neptune (called ice giants) and like Jupiter and Saturn (called gas planets). The discovery of the solar system analog could not have been accomplished without amateurs, particularly in New Zealand, where there were no professionals in MicroFUN.

Amateurs in New Zealand and other countries in the Southern Hemisphere are particularly valuable to astronomy, though to hear Gould tell it, you might not reach that conclusion. When McCormick first sent him data, Gould's expectations were low. "Have you been there [New Zealand]?" he asks me, rhetorically. "It's really green. Green pastures everywhere." While that sounds idyllic to everyone else on Earth, to an astronomer like Gould, green pastures are a sign of everything wrong in this world. And the tone of his voice when he says "green pastures" makes the condemnation apparent. But before I can connect the dots, he laid it out, "It rains often. They rarely get a good view of the sky."

In Gould's paper summarizing planetary discoveries from microlensing, he mentions his initial skepticism about McCormick: "Jennie McCormick, a New Zealand amateur, sent me an email one day saying 'I have data on your event, what do you want me to do with it?' Of course, it seemed preposterous that a 10" telescope in one of the wettest places in world could make a material contribution, but I started sending her our microlensing alerts. She contacted Grant Christie, another NZ amateur, who ultimately made contact with almost a dozen other amateurs around the southern hemisphere."

McCormick remembers collecting those data. On that night, she was initially planning to observe variable stars selected by the professionals in the CBA. The stars on that night's menu were low on the horizon. Before her astronomy hobby had exploded like a supernova, McCormick had planted a palm tree on her property. That night the leafy fronds blocked her view of one of the focal stars. She tried the other object but it was obscured by her TV antenna. She would soon sacrifice both, but for that evening she was itching to observe something useful. And so the palm tree, TV antenna, and her curiosity propelled her to try MicroFUN. McCormick remembers her notorious e-mail and how she later felt embarrassed by her casual tone because she thought she was corresponding with a student from Ohio State University but, she says, "Little did I know he was Professor Gould from Ohio!"

With its dark and stormy nights, New Zealand's weather creates the perfect setting for novels, but these conditions do not entice investment in large observatories and, consequently, professional astronomers are few and far between. The scarcity of professionals makes the New Zealand amateurs much more valuable, particularly for astronomical events like microlensing that require observations that last longer than the duration of one night. At any one point in time, half of the earth is in the dark, with the ability to see in only one direction. As the earth rotates, people on the other side can take over observations, like the passing of the baton from one observatory to another. Together such a global network of amateurs and professionals can construct more complete observations, like those leading to the discovery of planets.

Gould recognizes the importance of amateurs in the discovery of planets, asteroids, supernovas, and more, and even in the founding of the entire method of using microlensing networks to find planets. Because the amateurs have day jobs and less powerful telescopes, Gould wanted to optimize their time, so he limited the MicroFUN requests to high-magnification events. He soon realized that all in the network—even those with access to professional-class telescopes—were most efficient when also limiting their time to high-magnification events. Now he has used the proof of concept to obtain funding to create a microlensing network of telescopes that view sixteen square degrees of the sky. Amateurs don't (yet) have telescopes with these bigger fields of view, so they will be left out.

Gould mentions several amateurs in MicroFUN and explains to me in detail each of their most notable discoveries, revealing that he knows their individual contributions as well as he knows those of his professional colleagues. But he makes it clear to me that he has never (and will never) devise programs explicitly for the purpose of including amateurs. With MicroFUN, he had created a network of professionals, and amateurs were unexpectedly good enough to join in. On the other hand, many citizen science projects actively recruit participants, train them, and incentivize their continued participation. Gould never trains the amateurs or cultivates their participation. He explains, "We design program to be at forefront of science. Period." If amateurs are dedicated enough, they find a way to contribute, but he isn't going to hold back on using new technologies even if it makes amateurs' contributions obsolete. Besides, he is certain that amateurs will adapt and find new ways to contribute to science. McCormick knows that amateurs like herself are respected. "American professionals are great to work with," she says. "They value what amateurs do. They rely on us. When they ask for data, they get it."

New telescopes are lessening the need for amateurs in microlensing events—and in other areas too. The primary contribution amateurs have made to the field of astronomy over the years has been discovering supernovas, stellar explosions that mark the end of a star's existence. Gould refers to supernova hunters as "quite an industry of amateurs." For example, the Reverend Robert Evans has found thirty-two supernovas, a record for visual discoveries of this stellar phenomenon. (One has to know the night sky well enough to recognize a bright star that is not on the star charts.) Once a supernova is found by amateurs, professionals get more detailed data from bigger telescopes to answer cosmological questions; supernovas serve as standards for measuring the expansion history of the universe. So far, cosmologists trace expansion back to nine billion years. Studies of supernovas led to the discovery of dark energy, which astronomers think makes up 72 percent of the universe (4 percent ordinary matter and 24 percent dark matter make up the rest), unless Einstein's theory goes the way of Newton's due to some as-yet-to-be-seen flaw in our understanding of gravitation. According to Gould, amateurs topped the charts in finding the most supernovas close to Earth until last year. Now there are professionally driven projects that are automated and geared toward looking at the whole sky, which will find supernovas out of the reach of amateurs—especially very, very distant ones halfway across the universe.

McCormick notes the same possibility when it comes to asteroid hunting, an area previously the domain of amateurs. Astronomy can appear to be an esoteric science, one that attracts those with existential crises about whether we are alone in the universe. Yet when research is not focused on finding extraterrestrial life, it is often focused on saving our own. Astronomers estimate that a large object might slam into the earth every hundred million years, and so NASA wants to be prepared to stop the next one. NASA astronomers can detect and track near-earth objects that are as small as one kilometer wide. That's comforting, because geological evidence suggests that it was an object ten kilometers wide that struck the earth sixty-five million years ago, causing an explosion that altered the climate and thus wiped out the dinosaurs. But, by some estimates, something as small as 140 meters might put our future at risk. Astronomers have sifted through almost a quarter of the sky, finding over thirteen thousand objects in our solar system, including over three hundred asteroids and over sixty comets. NASA ran an automated telescope at the Catalina Observatory in Arizona that scanned the sky in order to detect asteroids. With computer algorithms, they can detect 90 percent of asteroids on the recorded images. They engage citizen scientists online to find the remaining 10 percent by searching images in Asteroid Zoo, similar to Galaxy Zoo. NASA designed the automated telescope, combined with crowdsourcing volunteers, to be a completely thorough asteroid detection system, making amateurs and their small telescopes obsolete for asteroid hunting.

Yet McCormick found something NASA missed: an asteroid 250 million kilometers from the sun. Asteroid 2009sa1 is validation that people still outperform technology; a 2009 headline read, "Kiwi Spots Asteroid before NASA." NASA has designated McCormick's discovery 386622 and assigned her the naming rights. She has ten years to select a name, and there are a few simple naming rules: she isn't allowed to name it after herself, her pets, or military leaders who haven't been dead longer than a hundred years. McCormick feels the weight of this responsibility because she knows it was a fluke and that it is unlikely that she, or other amateurs, will continue to find asteroids because, as she puts it, "Now the big survey satellites are hoovering up everything of interest."

With the majority of humanity in the Northern Hemisphere, most images of earth show the northern continents. Celestially speaking, the Southern Hemisphere is actually on top because it has a better view of our galaxy. Australians, New Zealanders, and South Africans have the best view of the center of our galaxy, the Galactic Bulge, which stretches east to west directly over the Southern Hemisphere. If you stand closer to the equator, or farther north, the Bulge is at the horizon and much more difficult to view.

We experience the landscape we're present in, whether forest, field, or green pasture. Astronomers, on the other hand, experience the Earth's landscape as a place to stand, a vantage point from which to view the even larger universe. They stand away from lights, and on dry, elevated land where they wheeze in the thin atmosphere, to get a less-obstructed view of space.

On a clear night people should be able to see thousands of stars. Instead, the majority of earthlings can only see a handful of the brightest ones. Most people think of the Orion constellation as a few points of light connected by imaginary lines, but there about 250 stars in Orion that are visible to the naked eye—that is, when there's no light pollution. Ninety-nine percent of people in the United States and

Europe see only light-polluted skies, and over 60 percent of humans live in places where they cannot see the Milky Way. Out of sight, out of mind.

Globe at Night is a citizen science project in which people measure light pollution and, by doing so, begin to realize what they are missing. It is one of the few citizen science efforts where people go outside at night. The project involves observing constellations in the night sky while the observer adds no artificial light (no flashlight or outdoor lights). Participants estimate the neighboring sky glow based on the amount of stars in a particular constellation that are visible to them. If sky glow is strong, then few stars are visible.

Light pollution has grown because light has become so cheap that we throw much of it away, tossing it toward the sky. William Nordhaus, an economist at Yale University, estimated that a Babylonian citizen would have worked forty-one hours to buy enough lamp oil to equal a seventy-five-watt bulb illuminated for one hour. A colonial American during the Revolutionary War would have worked five hours to buy that equivalent in candles. By 1992, a typical American using a compact fluorescent bulb could earn that amount of light in less than one second. As lighting gets cheaper, we use more of it. This is called the Jevons paradox, where technological improvements are counterproductive if the resultant savings are spent rather than saved. On the whole, we humans are not banking the savings from increased efficiency in generating electricity and bulb design—we just use it, even waste it. All the gains in energy efficient lighting are lost by a corresponding increase in demand and overillumination of the planet. Though the price of light is lower and lower, the health and well-being costs are higher and higher.

Before lighting, the world faced darkness at the end of every day, a darkness full of hazards, real and imagined. Street lighting brings feelings of safety, but excess lighting blinds us to where we are situated in the universe. Could there one day be darkness at the end of the tunnel as we realize the importance of being able to view the starry skies? Can we ration our use of lights, even though they are cheap? Towns around Lowell Observatory and Kitt Peak Observatory in Arizona have passed local ordinances that mandate a restriction on the total lumens per acre and place curfews on outdoor illumination. There are dark sky parks where night visitors can view the Milky Way. When species go extinct, they are lost forever. Fortunately, it is possible to restore the lost night sky.

Amateur astronomy and seeing the enormity of stars in the night sky changed McCormick's view on life: "When you understand the night sky and how immense it is, then you realize how insignificant you are. We are on a little blue planet around one star in a universe that contains zillions of stars." She notes, in awe, "The universe is beautiful. The night sky is so amazing and wonderful," and then adds, matter-of-factly, "I'm a total atheist. We are all made of star dust. The sun made us all."

Yet McCormick finds optimism in astronomy. Her efforts could give earthlings the greatest gift ever: company. McCormick explains, "We proved that stars are homes to planets just like our solar system. We found a small solar system analog and it's probably got other planets that we can't yet detect. I hope we can one day find an earthlike planet. I'm sure there is life out there. It's just a matter of saying hello."

Nature nerds and science fiction fans are not the only ones interested in science. The term *fan*, short for *fanatic*, originally referred to those enthusiastic for a particular sports team. Then fandoms formed to unite people with common interests, whether those interests were in celebrities, books, movies, or entire genres. Fans form fan clubs, create fan art, write fanzines, and meet online and at conventions. Do you miss Harry Potter? There is plenty of Harry Potter fan fiction to enjoy. According to Clay Shirky, author of *Cognitive Surplus*, in the Internet era people no longer simply act as consumers of content, but rather share their creativity and imagination through fandom. Citizen science is fan science, a collection of activities in which people go further than crafting funny videos of cats. When people want to dedicate their creativity, enthusiasm, and time into something meaningful, they take up citizen science. As we take a look at those carrying out citizen science entirely online, we'll see surprising types of fans.

PART 2
The Necessity of Leisure

THE COMPUTER GAME POKÉMON GO MANEUVERS AN ASTOUNDING number of people into exploring their neighborhoods, having fun by finding virtual reality creatures detectable only through their smartphone screens. It shouldn't thus be a big leap to have fun by finding and observing real wildlife and other phenomena with the naked eye, binoculars, or telescopes, as we've seen in part 1. It also isn't a big leap to biochemistry and microbiology, fields in which citizen scientists study phenomena invisible to the naked eye. Nor is it a big leap from virtual reality games to conservation biology, where the eyes of citizen scientists are open to new perspectives, because, as Henry David Thoreau said, "It's not what you look at that matters, it's what you see." In the following chapters we'll examine these newer disciplines and see that collaborations between scientists and the public takes on new and unique formats.

CHAPTER 5

Biochemistry
Protein Folding Is Magic

Play is often talked about as if it were a relief from serious learning. But for children, play is serious learning.

—FRED ROGERS, AKA MR. ROGERS

ON OCTOBER 10, 2010, THE FIRST EPISODE OF *MY LITTLE PONY: FRIEND-ship Is Magic* aired. Sebastian Lamerichs was then a fourteen-year-old student at the Australian Science and Mathematics School in South Australia. His online world of forum boards was soon populated by avatars of ponies and unicorns with coiffed manes. The fandom bloomed like acne on an unwashed pubescent face; he noticed students donning My Little Pony shirts, and a few of his friends started watching the animated series. Lamerichs was shepherded into the herd, "lured by the offer of free pizza" at a winter wrap-up party for the season finale. When the sun rose, he had already borrowed and watched the entire first season. Overnight he had become a Brony, a male ("bro") fan of a rainbow-festooned television show marketed to little girls.[50] His next step on this enigmatic hoof path was to become a member of Brony@home, the largest citizen science team based in fandom.[51]

---

[50] There are female Bronies too, though they are also known as Pegasisters.

[51] Lamerichs writes, "While we certainly weren't the very first fandom-based distributed computing team around, all of the others I've seen were never really successful (5–6 users at most). We have had two charity competitions on Reddit in the past where other fandoms (The Last Airbender, Adventure Time, and Harry Potter) had started up distributed computing teams, but the MLP [My Little Pony] teams won by an order of magnitude on both occasions—despite the MLP subreddit having less than half as many subscribers. We're definitely easily the most active fandom in this aspect."

Brony@home is a distributed computing team. Some scientific problems are so computationally intense that they require more computing juice than can be squeezed from a single computer. To meet the computational demands of these special problems, researchers use the Internet to tap the unused central processing unit (CPU) and graphics processing unit (GPU) cycles of thousands of personal computers distributed around the globe. The phenomenon of weaving webs of personal computers into high-powered distributed computing systems began with a program called SETI@home (in which SETI stands for "the search for extraterrestrial intelligence"). Participants simply download a software program, and allow their personal computer to be accessed as part of a distributed computing network. The success of SETI@home (success being defined by a highly amplified search, not by finding extraterrestrial intelligence) was followed by many distributed computing projects—particularly in astronomy—with @home in their name, such as Asteroids@home, Einstein@home, and Stardust@home. These can be found on the largest hub of distributed computing networks: BOINC, which stands for the Berkeley Open Infrastructure for Network Computing. Thanks to volunteers who link their personal computers to the network, researchers in a variety of disciplines, now with a strong showing in biochemistry, are able to crunch otherwise uncrunchable numbers with enormous processing power by tapping tiny bits of unused computer cycles dispersed across the world. Such a network has significantly more horsepower (or pony power?) than any individual computer ever could.

That's saying something given how powerful personal computers are today compared to just a few decades ago. In the 1950s there were only a few hundred computers in the world, each as big as a warehouse. By the 1960s smaller transistors allowed for smaller computers, like mainframes the size of bank vaults. By the mid-1960s, the microchip made it possible to have personal computers, but they were still too expensive for most people, and they weren't cost-effective for businesses. The computing power of a CPU was proportional to the square of its price. That meant you could double your financial investment and get four times the computing performance for your purchase, something known as Grosch's law. Big mainframes were the economical choice for businesses until 1975, when Bill Gates started Microsoft and Steve Jobs started Apple. Grosch's law could not rule the world of Silicon Valley; today a single, cheap CPU is faster than the gargantuan mainframes of the 1960s. Investing more money no longer results in a faster CPU, at least within the range of uses for the average person. To get the best bang for your buck, you must string together thousands of small CPUs. Today the world's fastest computers are computer *systems*.

As we saw in Part I, those with natural history hobbies easily fit the citizen science mold. The hobby involves making observations, and the hobbyist enjoys sharing these observations for scientific purposes. It's a win-win scenario. In an analogous way, when someone enjoys high-performance computers, the hobby, known as overclocking, involves connecting one's hardware to networks as a way of competing, solo or on a team, and the hobbyists are glad these networks advance scientific purposes. In order to understand overclockers as citizen scientists, let's take a deeper look into the field of biochemistry, where distributed computing is necessary to make breakthroughs in medical research.

In biochemistry, research related to the structure of proteins is challenging because proteins are very small. Their composition is studied by extracting DNA from living cells (in vivo). Their structure is sometimes viewed with specialized, expensive equipment in laboratories (in vitro). The frontiers of the field involve studying the structural configuration of proteins with computer models (in silico). It's computationally intensive, and hence entire networks of computers distributed geographically across the globe support Folding@home, the distributed computing network that initially attracted the Brony@home team. Folding@home is a racehorse network with over 100,000 computers creating a cumulative speed of forty petaFLOPS,[52] which is faster than *all* the BOINC projects

---

[52] FLOPS (floating-point operations per second) is a measure of computer performance; petaFLOPS are FLOPS of 10^{15}, a scale far bigger than kiloFLOPS (10^3), megaFLOPS (10^6), gigaFLOPS (10^9), or teraFLOPS (10^{12}) but not as big as yottaFLOPS (10^{24}).

combined. A project with a similar purpose is Rosetta@home, which is hosted by BOINC, and it operates at over 110 tcraFLOPS. Neither project would be possible without thousands of volunteers linking their computers to donate the otherwise idle time of their CPUs and GPUs.

The research priority of Folding@home is related to protein folding. Proteins are the spark plug for nearly all reactions in living cells and control virtually all cellular processes. Protein function is related to their many configurations: crystalline proteins are in the lenses of eyes; collagenous proteins are in skin; fibrous proteins are in hair; enzymes are proteins that assist chemical reactions. The configuration—or fold—of a protein is essential to its function. All types of proteins can be damaged simply by changes to their configuration, even if their composition stays the same.

Stress, age, heat, and cold are a few conditions that can damage proteins, and damaged proteins are referred to as denatured. Although proteins are too small for us to see with the naked eye, we can see evidence of denatured proteins by cooking an egg. The proteins in the egg's albumen are loose, translucent globs when raw, but clump together as a tight white mass when denatured by heat. Proteins don't function when denatured. If any protein misfolds and does not configure into the precisely right structure, it won't function properly either.

Understanding protein folding increases the odds of finding treatments for Alzheimer's disease, cystic fibrosis, AIDS, mad cow disease, and emphysema, to name a few. All told, about twenty diseases, which manifest in seemingly unrelated ways, have a common cause related to protein folding or misfolding. Each of these diseases arises in one of two ways. One way is when misfolding creates proteins that won't carry out their essential functions. An example is cystic fibrosis, which is characterized by a tiny nick in the protein (that is, the loss of one amino acid from a sequence), and this in turn causes consistent misfolding of a protein that regulates membrane transport (the protein is called a cystic fibrosis transmembrane conductance regulator). With no properly folded proteins for this particular transport of chloride ions, the disease manifests in cells that produce mucus, sweat, and digestive fluids. Instead of producing thin and slippery secretions that lubricate, cells in people with cystic fibrosis produce thick and sticky secretions that block tubes and passageways. The other way diseases arise is when misfolded proteins become toxic to cells because they aggregate into masses that interfere with other cell functions. When misfolded proteins gather into molecular wastelands, victims have diseases such as Alzheimer's, Parkinson's, and Huntington's.

Because the function of a protein is determined by its folded structure, not simply by what it is made of, a misfolded protein's function can be restored by folding it properly. Some changes to protein configuration are indeed reversible. Called renaturation, the ability of proteins to refold properly gives great hope to the possibility of treating diseases caused by initial misfolding. Even the proteins in cooked egg whites can be renatured to their correctly folded configuration so that the albumen is restored to its translucent liquid state. Hervé This and Nicholas Kurti, leaders in the science of culinary phenomena, which they call molecular gastronomy, found that egg white proteins can be renatured by adding vitamin C or sodium borohydride to the whites for about three hours.

What these chefs did was not too different from what led Christian Anfinsen to be awarded the Nobel Prize in Chemistry in 1972. In a series of experiments during the 1960s, Anfinsen heated a protein called ribonuclease that he extracted from the pancreatic tissue of a cow. When heated (a process that involves adding energy) it no longer functioned because it unfolded. When it cooled (a process that involves releasing energy), it refolded and functioned properly again. He concluded that when this particular protein enzyme (which comprised 124 amino acids) was denatured by heat, it maintained its sequence of amino acids and refolded back into its precise form. He demonstrated that the folded state is the lowest energy state. Research for potential cures for many diseases is now based on the premise that folding occurs along the path of least resistance.

The *in silico* study of protein folding thwart the powers of individual computers because it is computationally intensive to find the path of least resistance along which a protein will fold. Imagine modeling every unique snowflake. Even though the composition of each snowflake simply includes two hydrogen atoms for every one oxygen atom, the unique configurations possible are endless. Complex structures can arise from simple compositions. Protein composition is a sequence of amino acids dictated by DNA blueprints. Other cellular organelles, knowns as molecular chaperones, guard each newly printed sequence to keep it from folding before the entire sequences is in place. Once the linear chain is complete, the chaperones let go, like releasing a stretched-out Slinky, and snap, bend, and twist; the chain will collapse into a three-dimensional functional protein. Proteins achieve their three-dimensional shape by folding like origami paper. The right shape is biologically useful. Even though there is a finite number of amino acid combinations in linear sequence, there is an astounding number of possible three-dimensional configurations.

In the late 1960s Cyrus Levinthal tried to pin down the number of possible configurations for a protein. Even with conservative assumptions, for example, that a protein had only one hundred amino acids and could only take two different spatial orientations, there were theoretically 3^{300} or even 10^{143} possibilities—the possibilities are nearly endless. Even if a protein could try one hundred billion different folds per second, it would still take far too many years to go through all the possibilities. The properties of the amino acids, and how they bond together, influence how the protein will fold. The avenue of least resistance is similar to how having a piece of paper that has been creased previously makes it easier to fold along that crease again.

Even though folding happens quickly, to sequentially go through all the almost endless possible folding pathways is computationally intensive, even with the rule of following low-energy states. The speed at which proteins fold actually makes the greater roadrunner's eye blinking appear as slow as a herd of asthmatic turtles moving uphill. Folding happens as fast as one millionth of a second. Many proteins fold on a timescale of one million nanoseconds. For some reason (either pony magic or the snowfall of endless possible configurations), simulating the fraction of a nanosecond takes a computer a long time. In one day, a computer can simulate fifty nanoseconds. According to the Folding@home website, it would take twenty thousand days (almost fifty-five years) to simulate on one computer the amount of folding occurring in one day on the network of approximately 500,000 processor cores.

The Brony@home team was created in April 2011 by an Australian with the online name Hiiragan. Hiiragan was nineteen years old—five years older than Sebastian Lamerichs—when the pony images and memes began dashing into his online forums. He recalls, "Frankly, it was rather annoying to see countless threads with plenty of potential discussion material to be ruined by people making random and usually unrelated posts [about ponies]." He checked out discussion threads on 4Chan, an image-sharing website where many users strongly resented the pony stampede.[53] Bronies bucked 4Chan and moved on to create PonyChan, an image board just for content related to My Little Pony. The controversy piqued Hiiragan's curiosity and soon he trotted into the Brony coral.

Hiiragan had been hooking up his personal computer to distributed computing networks since the age of thirteen. Back then, he explains, his computer was "a Pentium 4 with 256 megs of RAM." Though small on its own, he knew "a few thousand P4s worldwide would put a noticeable dent into some work units," so he joined a distributed computing team. His team started petering out, entirely coincidentally, soon after he became a Brony. "I decided to jump ship. Or in my case, build a new ship," he told me via e-mail. The Brony team started out with a few volunteers, but grew fast. With a website to attract new teammates and by spreading the word across sites related to My Little Pony, the Brony@home team galloped through a few events where teams competed to donate the most CPUs to a particular research question. Soon, as Hiiragan relayed, "our member count exploded across not just Folding@home, but BOINC projects as well, which we eventually expanded out into." Brony@home now has over a thousand members and

---

[53] For a while, 4Chan banned My Little Pony images.

ranks in the top fifty for donated CPU and GPU cycles out of thousands of teams in BOINC.

Like most citizen scientists, Lamerichs participates in distributed computing networks because he likes the science and the sense of community effort. Unlike other citizen scientists we've encountered so far, Lamerichs is highly driven by statistics and the competition they create. A sizable segment of online citizen science is gamified: researchers have added elements of games to nongame situations: people enjoy online tasks where they can earn immediate points, accumulate points to earn badges, progress from one level to another as evidence of gaining skills, gain recognition on leaderboards, and even immerse their scientific contributions in virtual stories. A new class of social scientists, those who study human-computer interactions, examine how well these elements work to produce desired behavioral outcomes. (In this case, the desired outcome is continued sharing of unused CPUs and GPUs, but in other contexts, the desired outcome could be commerce, education, health or exercise, work, innovation, or data gathering.) According to ethicists, online game elements are considered a persuasive technology because they hold the potential to alter user behavior by influencing people in overt, and covert, ways. When someone is motivated to "earn points" and to win, the owner of the game has a lot of influence on the player. And vice versa: when the owner of a game relies on thousands of players, the players collectively can influence the owner. When the Folding@home owners decided to change the point system, those donating their computers to the network cried foul. Ever since, how points are allocated has been through a mutually agreed upon system for fair competition.

Lamerichs feels the competition is key: "It's one of the greatest reasons for the success of our team. We got to number one globally in Rosetta@home by points per day. Some people do seem to be of the opinion that competition is somehow inherently bad, but without that motive, there wouldn't be anywhere near as much work done."

Compared to citizen science projects we've encountered so far, participating in distributed computing might appear to require the least effort: participants can literally eat pizza and watch My Little Pony while their computers chug away. Why does it require the most carefully crafted incentives to recruit and sustain participation?

For answers I look to Vickie Curtis, whom I met a few years back at a citizen cyberscience conference. Curtis, who studied online citizen science for her dissertation at the Open University in England, explains that there are two types of participants in distributed computing networks. Some are ordinary people like her, with laptops or PCs who hear about an @home project and take a few minutes to enroll their computer. Such casual users aren't motivated by the statistics, because they will never make it onto a leaderboard. Curtis says that her laptop would take weeks or months to gain thousands of points, something an overclocker could earn in a day.

Overclockers are the second type of participants;[54] they invest tremendously in their equipment for distributed computing. Most in distributed computing tend to be men,[55] who custom-build their own computers to process the maximum amount of data possible without burning up (quite literally; computers can overheat). They mainly rely on GPUs (the powerful processors used for video game graphics) because they are more powerful than CPUs. They invest thousands of dollars in building "folding rigs" and prevent their folding rigs from overheating by building sophisticated cooling systems, sometimes with liquid nitrogen. Instead of drag-racing customized cars along an abandoned canal, overclockers use Folding@home as a place to compete in the performance of their customized computer systems.

Overclockers are the main fuel of distributed computing networks. To entice overclockers to join distributed computing networks, projects will award points and display the screen names of top contributors on leaderboards. Some also hold regular competitions to see which individuals

---

[54] There are also "extreme overclockers" who hold conventions and hard-core competitions. When Curtis requested interviews of Folding@home participants, about 90 percent of respondents were overclockers.

[55] Among the respondents to Curtis's request for interviews with Folding@home participants, less than 2 percent were female, and those few are predominantly working in information technology or electrical engineering fields.

and/or teams can process the most data, such as in Folding@home's annual Chimp Challenge. Because one individual can only own a limited number of computers, participants band together into teams in order to increase their odds of reaching the top of the charts. Some join groups to learn more about overclocking from other participants. Many form friendships online.

Projects like Folding@home post participants' scores every hour, on the hour. Researchers designed the system such that points incentivize participants to optimize their computers in ways that will increase the scientific gains of the project. When Folding@home sends a project to the network, it also runs it on a computer that serves as a benchmark. Points are calculated by comparing the progress of a benchmark computer to each of those in the network, plus every computer earns a base value of points for participating. The benchmark processor is an Intel Core i5 CPU 750 at 2.67 gigahertz with a Linux operating system. Members of the Brony@home team are fluent in these types of details about the processor type, its speed, the drive size and speed. One Brony told me that points didn't matter when he was a teenager, but he always looked forward to getting a higher performing machine, like an I-5760 quad core GTX 680 with eight graphics cards . . . you get the idea: overclockers love data, statistics, and specifications. The Brony@home team is populated by individuals participating in Brony fandom, some of whom are overclockers. Together this herd has been a top contributor to Folding@home.[56]

Through interviews with people in Folding@home, Curtis found that a quarter of participants don't feel that they are citizen scientists involved in scientific research. They feel more similar to those who donate money to research than those who carry out research. Curtis attributes this feeling to rhetoric around citizen scientists. The Folding@home creators refer to participants as "donors." Other online citizen science projects, like those in Zooniverse (see chapter 4), refer to participants as "collaborators" and acknowledges them in publications, often with

---

[56] The Folding@home lab has published 114 scientific research articles since the project's launch in October 2000.

coauthorship credits.[57] Whether or not they self-identify as citizen scientists, they have found a way to apply their hobby toward a meaningful cause. They are motivated by more than points: Many have family and friends battling diseases studied by distributed computing; some have lost family members to these illnesses. Some overclocker forums on Folding@home are virtual shrines to loved ones lost to Alzheimer's, Parkinson's, and cancer.

Folding@home and Rosetta@home have complementary research objectives related to the protein structure. To citizen scientists, the key difference is that Rosetta@home emphasizes the design of proteins and the link between the folded structure of a protein and its function, while Folding@home is focused exclusively on modeling how proteins attain their structure through folding. To the overclocker, the important difference is that Rosetta@home is a BOINC project and Folding@home is not; plus, they differ in various technical specifications. For some participants the main difference between Rosetta@home and Folding@home is the design of the screensaver, which is an animation video of a three-dimensional jigsaw puzzle in action, showing the different shapes the computer is testing. For Folding@home, the protein on the screensaver is made of round beads; for Rosetta@home, the protein looks like ribbons and thread.

David Baker, the lead biochemist on the Rosetta@home project, speaks to me from his research lab at the University of Washington.[58] He explains that people were captivated by the Rosetta@home screensaver. The computer made it look easy, too passive, and many people came to the conclusion that they could do a better job at solving these three-dimensional puzzles than their computers. They spoke

---

[57] Participants in online transcription projects run by the Smithsonian Institution call themselves "volunpeers."

[58] It is a small-world coincidence that David Baker and Lucy Fortson, who launched Zooniverse (see chapter 4), were classmates at Garfield High School in Seattle, both have parents who were physics professors, and both got involved in citizen science around the same time.

up and told Baker and his team that they wanted to guide the folding on their computers. To his credit, Baker believed in their abilities. Game on.

Baker teamed up with computer scientist Seth Cooper. Together they created Foldit, which launched in 2008. In Foldit, individuals, sometimes working as teams, compete in trying to solve protein-folding puzzles. Participants—now called players—are donating their mental powers rather than their computer powers. As with distributed computing, Foldit players associate with teams. High scores are based on biological principles. They twist, tuck, and tug protein backbones and side chains to find the most stable, low-energy configuration. A higher score indicates a higher likelihood that the protein shape is the lowest energy state. The computer continues to do the low-level number crunching while people use their spatial reasoning ability.

As proof of concept, the Foldit group pitted players against computers in solving an already known protein fold. The players outperformed the computers. As Cooper puts it, "The computer gets stuck, and people use intuition to get unstuck." As if navigating a maze of dead-end streets, people think of creative ways around a problem, and computers don't have creativity. The solution requires human imagination.

Next, the Foldit team gave players a real problem that had stumped computational methods for ten years. Their charge was to figure out the folded shape of a special type of protein associated with AIDS in monkeys.[59] Of the 300,000 people who had registered at the time, two teams figured out the most likely fold of the protein. These teams, Void Crusher and the Contenders Group, were coauthors on a paper with lead researcher Firas Khatib and other colleagues in 2011 in the journal *Nature Structural and Molecular Biology*.

As we saw in chapter 2 with eBird, citizen science projects have many more registered participants than active participants. Curtis no-

---

[59] The Mason-Pfizer monkey virus originated in the breast tumor tissues of rhesus macaques in 1970. It was thought it could be used for cancer research, but when newborn rhesus macaques were inoculated with it, they developed a wasting disease accompanied by opportunistic infections like pneumonia and rashes.

ticed that when scientists refer to the popularity of the projects, they "typically refer to *registered* participants, as opposed to *active* participants, because the expectation is that the public will judge project success based on the number of participants, even when scientists judge it successful based on the number of new discoveries." While there are hundreds of thousands of registered Foldit players, only two to three hundred actively attempt to solve most of the puzzles. The overwhelming majority are filtered out by the lengthy tutorial puzzles.[60] Of the active players, almost 80 percent are male, and almost seventy percent of those are over the age of forty; most of them don't play other computer games. From this small percentage, twenty to thirty are the core who discuss the game on forums, dominate in-game player statistics, write content for the game wiki, and mentor new players. They log in many hours: in one survey, many participants had been playing for more than two years, spending about fifteen hours per week at the game. As one player put it, "the real point is that Foldit simply allows us folks without the proper CVs, and [who] would crawl over broken glass to participate given half the chance, an opportunity to do this stuff. It's that simple." When social scientists asked players about the skill set needed to be a player, most responded with personal or character attributes, such as perseverance, patience, dedication, people skills, and determination, as well as obsessiveness. One respondent referred to these traits as "the right stuff." Another player put it more succinctly: "By far the two attributes that help Foldit players are an obsessive personality and scientific inquisitiveness."

Foldit has not stopped with puzzles of protein folding. It has grown into an experiment involving players, scientists, and game designers so that the game continues to evolve with new visualizations, tools, and

---

[60] Before players can compete in solving a protein puzzle, they need to complete a series of thirty-two puzzles (already solved) as part of a tutorial. It seems like an impossibly high bar to set when recruiting volunteers.

other elements to make the game more effective. Researchers added techniques called machine learning in which the thousands of strategies developed by players are taught to the computer and to other players. Over the past five years, Foldit has automated some puzzle moves. Players can use a coding language called Lua to create sequences of moves called recipes, and players share recipes with their team and the entire Foldit playing community. Each recipe is geared toward a particular aspect of folding, to solve a particular problem in the configuration. According to Curtis, "some players eschew these recipes and prefer to play by 'hand.'" They are known as "hand folders," and Curtis says they talk about playing "by instinct." The Foldit developers are aware of the preferences of the subcommunity of hand folders and accommodate them by releasing some puzzles that cannot yet be solved with recipes.

The long-held principle that two minds are better than one is now illustrated with online games like Foldit and has the backing of scientific studies that label the phenomenon collective intelligence. Psychologists in the early 1900s defined intelligence as a trait of an individual. Someone with high intelligence, or what psychologist Charles Spearman in 1904 called the g factor (for general intelligence),[61] will perform consistently high on different cognitive exercises. It has been repeatedly found that if people do well on one mental task, they tend to do well on other tasks. Such cognitive ability predicts many aspects of one's future—grades in school, success at different occupations, and even life expectancy. Spearman was influenced by the statistics developed earlier by Francis Galton, who wrote about crowd intelligence. Galton watched as people at a carnival guessed the weight of an ox. Galton recorded each person's guess. No one was exactly right, but when he took the average of their guesses, he obtained the correct answer. Is there a collective wisdom that enables groups to figure out problems better than individuals?

---

[61] Though a psychologist, Spearman excelled in statistics and his namesake—the Spearman's rank correlation coefficient, or Spearman's rho—is used to describe the degree of dependence between two variables according to their ranked relationship.

In 2010 Anita Woolley and her colleagues published a paper in *Science* demonstrating that yes, collective intelligence—what she coined the *c* factor—does exist. First, they found a positive correlation on performance across tasks for groups (that is, if group A performed well at checkers, it also performed well at negotiating). Second, the variation was largely explained by a specific combination of characteristics about the group: social sensitivity, the ability to take turns, and a combination of genders. Third, group intelligence was not related to the average, nor the maximum, intelligence of individuals in the group. We may not be able to give significant boosts to individual intelligence, but the existence of collective intelligence means we can potentially design groups with the right combination of people to achieve high collective intelligence.

Woolley and her colleagues honed in on the characteristics of highly intelligent groups. In total, about seven hundred people were involved in the study, but each group comprised only two to five individuals. If you volunteered, you were run through a battery of mental tests, first alone and then with a group. You might start with brainstorming tasks that demonstrated creativity, and then move onto demonstrate your decision-making ability in tasks related to making moral judgments. Then you'd play checkers against a computer to test your performance and psychomotor skills. Finally, you'd have fun with visual puzzles to see how you did in contests and competitive tasks. By having individuals do the tasks alone, Woolley was able to classify each individual into a well-established taxonomy of intelligence types. She then assembled groups that contained individuals with a mix of types, and then measured the collective intelligence of each group.

As mentioned above, collective intelligence is unrelated to the average or maximum intelligence of individuals that compose a group. Nor did group cohesion, motivation, or satisfaction matter. Instead, what mattered for collective intelligence was the average social sensitivity of the group members, equality in taking turns to talk,[62] and the proportion of female participants in the group. Groups where a

---

[62] Such equality was measured by sociometric badges they wore.

few people dominated the conversation were less collectively intelligent than groups who passed the talking stick around equally. Groups with more female participants had higher collective intelligence, but this was because female participants tended to score higher on social sensitivity.

In Woolley's research, groups interacted in person. Researchers have assumed that group intelligence requires face-to-face interactions because one key factor, social sensitivity, is measured based on visual cues. But if face-to-face interactions are essential, then how could collective intelligence occur online, as among Foldit players? David Engel at MIT's Center for Collective Intelligence joined forces with Woolley to find out whether or not online interactions impair or support collective intelligence.

Engel and Woolley repeated the study, but with half the groups interacting online. The battery of exercises took about an hour, involving some parts you find in the Sunday paper, like unscrambling jumbled words and Sudoku, as well as exercises involving brainstorming, applying judgment, coordinating typing into a shared document, memory tests, and detection ability. The results were the same: online groups showed the exact same patterns of collective intelligence as in-person groups. Whatever factors are important in creating collective intelligence—even social sensitivity—can occur without communicating face-to-face.

It's no surprise, then, that highly successful Foldit teams are a good mix of people with different levels of mental prowess. Players sort themselves to focus on different parts of the puzzle. Some are good at just figuring out the beginning; others are good at the end stages, and will often complete puzzles that others begin but can't finish. According to Curtis, the finishers are called evolvers and they are highly skilled and top ranking. In the team she played on, there were three or four evolvers who would compete in finishing a puzzle, share their results, and then collaborate in finessing the final structure.

Alone or in groups, humans surpass computers in imagination, creativity, and spatial reasoning. Minds are superior to machines.

• • •

With people solving protein folding problems in Foldit, researchers were able to start using Rosetta@home in a new way: testing new design sequences. In this type of bioengineering, Rosetta@home processing works through algorithms to design new protein shapes that have never been known to exist, and then researchers try to create them in the laboratory. Could Rosetta@home design a protein that fights the flu or degrades plastic? Not long after Foldit became a distributed computing problem, its players wanted to design new proteins too. "The user community wanted to play God," Baker explains.

The first creation by Foldit players, their "Adam and Eve" protein, was not from scratch but involved redesigning a previously designed enzyme called Diels-Alderase so as to increase its ability to catalyze chemical reactions. A common stumbling block with designed enzymes is that they usually have a lower capacity to catalyze chemical reactions than natural ones. Foldit players accepted the challenge and their redesigned enzyme was more than eighteen times more efficient than other designs.

Scott Zaccanelli, known online as BootsMcGraw, is a Texan who works for a valve factory and moonlights as a personal message therapist. In his heyday he had been ranked in the top ten globally on Foldit and was part of the Contenders team. He was the first Foldit player to design a protein that Baker sent to the lab for synthesis. It was called fibronectin, and looked like the most promising design for a new antibody-like compound. It turned out not to be stable, but neither Baker nor Zaccanelli were discouraged. Science is trial and error. Zaccanelli had been playing Foldit almost every night for over five years. He didn't stop because of the scientific challenge but a social one: a falling out with another Foldit player.[63] Zaccanelli's departure illustrates a consideration that

---

[63] As Zaccanelli notes in the "About me" section of his Foldit profile, "When the management lends more validity to the rantings of a spoiled, malicious teenager over the actions of players with six years and thirty thousand hours of commitment, it's time to move on to something more worthy of my involvement."

social scientists have discovered about citizen science: when individuals join a project they are motivated by curiosity in science, but their sustained activity in the project for the long haul is motivated and heavily influenced by social factors.

A similar project, called EteRNA, allows players to design RNA. Like protein design, researchers can use computer programs to design RNA, but the cutting edge of this type of bioengineering relies on trial and error as well as imagination—the latter being an activity at which humans surpass machines. Over 37,000 people have registered to work (meaning play) at RNA design puzzles through EteRNA.[64] They earn points based on how well they perform certain tasks, and the reward is access to more tasks of increasing complexity. For a given puzzle, players eventually vote for the best design and the winning design is synthesized in a real lab. Thomas Rowles, senior executive editor at BioMed Central, interviewed some EteRNA gamers for his editorial in the journal *BMC Biochemistry*. As one gamer, going by the name WaterontheMoon, explains, "The difference between us and scientists (and there are scientists among us), is we have no idea what should work so we try everything." Ignorance is not only bliss but a gateway to creativity.

The abilities of the human mind over mainframes has spurred other online projects in biochemistry. For example, Phylo (short for phylogeny), an online citizen science game created by researchers at McGill University,[65] involves sets of mental challenges to align DNA sequences for comparative genomics. The results help identify the parts of the gene that may be related to certain diseases in a process called multiple sequence alignment. In Phylo, players slide colored blocks as if beads on an abacus, where each color corresponds to a nucleotide in a DNA sequence. They

---

[64] Like other projects, thousands register, but only a relatively small group of players are actively involved in contributing to the project.

[65] As we'll see in chapter 6, the Facebook game Fraxinus is similar, but focused on one genome.

slide the blocks until as many as possible align across rows. There are many regions of DNA that are not useful, but the segments that appear across species are those of shared evolutionary origin and therefore most likely the areas that are functionally important. Computational methods can do much of the alignment, but it still leaves small inaccuracies and refinements. As with Galaxy Zoo (see chapter 4), when Phylo launched in 2010 the university computer servers crashed, unable to handle the unexpected volume of thousands of simultaneous players. Once the servers could handle it, Phylo players quickly demonstrated that they aligned nucleotides better than a computer was able to do.

The Bronies in Rosetta@home and the players in Foldit illustrate that citizen scientists come in many varieties. Studies of participants and players have found motivations to enter the project grounded in the desire to help scientific research. That is the same primary motivation of on-the-ground citizen science participants, where their tools are binoculars or butterfly nets rather than personal computers. Studies have also found that sustained participation is generally grounded in the social aspects of the project (with Foldit, there is an Internet relay "chat" window with which players can communicate; also a forum, where they can direct-message one another). With Foldit, an additional factor is involved: players who stick with the project savor the challenge. Instead of merely pushing computing limits, as in Rosetta@home, they push their mental limits. In both projects, participants are simultaneously competitive and cooperative. In both, they form friendships. In one interview, a Foldit player referred to his team as his "folding family"; others talk about having made friends in the folding community. The degree to which players are fluent in the science varies quite a bit. One said, "I was diagnosed with relapsing-remitting MS about 4 years ago. So I am interested in medical research, particularly oligo-dendrocites—the cells that re-build the myelin." Another said, "I play by score; I have no real idea what I am doing, I just follow the score."

Online games for citizen science work because humans are great at spatial reasoning, creativity, and finding communities in which to form our own identities and connections. Our spatial reasoning includes pattern-recognition skills, excellent abilities with three-dimensional visualizations, and unparalleled ability to spot subtleties. We are creative enough to tackle a problem from multiple perspectives at once. We create communities, and fandom communities, based on common interests. Communications researcher Mikko Hautakansas at the University of Tampere in Finland studied Finnish Bronies, and confirmed that they are sincere in their enjoyment of the show even though it does not fit traditional notions of masculinity. Bronies have renegotiated male gender norms and constructed their own male identities. Similarly, Brony@home have renegotiated scientific norms and constructed their own way of engaging in science. They may be fans of fiction, but they are aficionados of solving real-world problems.

In earlier chapters we saw citizen science in disciplines with long traditions of relying on people with nature-related hobbies. By tapping into what people are already interested in, projects avoid many problems related to data quality. Good project design involves finding a good match with participant expertise. As Curtis notes, "It doesn't take very many people to make a difference to these projects; however, you have to cast your net far and wide in order to attract these small dedicated groups of participants."

Fandom groups like Whovians, Potterheads, and Trekkies tend to comprise tech-savvy youth. It was a pleasant surprise to find that the Bronies were an ideal match to distributed computing efforts. When participants and projects find one another, a good match translates into a job well done.

There are other disciplines like biochemistry where the subject matter is invisible and there is no existing base of hobbyists to attract to research projects. What happens when life invisible to the naked eye is made visible with citizen science, or when citizen science causes the walls that shield science in the ivory tower to become transparent? As we'll see in chapter 6, science *of* the people (and their microbes) becomes science *by* the people, which finally becomes science *for* the people.

CHAPTER 6

Microbiology
Invisible Worlds Go Public

This song is Copyrighted in U.S., under Seal of Copyright #154085, for a period of 28 years, and anybody caught singin' it without our permission, will be mighty good friends of ours, cause we don't give a darn. Publish it. Write it. Sing it. Swing to it. Yodel it. We wrote it, that's all we wanted to do.

—WOODY GUTHRIE (*message on mimeographed copies of lyrics distributed to fans in the 1930s*)

AFTER GRADUATING FROM GEORGIA TECH, WENDY BROWN RETIRED HER pom-poms from the Atlanta Falcons cheerleaders and moved to California to start her PhD studies in biomedical engineering at the University of Caliornia–Davis. In her spare time she stayed on the sidelines, first cheering for the Sacramento Kings in the NBA, and then synching her Rockette-style kicks with the Oakland Raiderettes. Brown is one of the Science Cheerleaders, a squad of three hundred current and former cheerleaders pursuing careers in science, technology, engineering, and math (STEM). Former Philadelphia 76ers cheerleader Darlene Cavalier founded the Science Cheerleaders in hopes that cartwheels will flip public stereotypes of scientists and cheer girls into STEM careers.

During the 2014 basketball season, Cavalier, Brown, and other Science Cheerleaders teamed up with the 76ers to create a citizen science halftime show. While rousing the fans with cheers and acrobatics, they spurred the crowds into action. Together with the 76ers Flight Squad (halftime entertainers), they blasted microbe collection kits (wrapped in T-shirts) out to fans with a powerful shirt cannon. They repeated their performance at major NFL, NBA, and MLB events. That spring, sports fans collected about four thousand samples of microbes living on their shoes and cell phones.

An invisible world of microbial organisms lives on us, in us, and around us. On every surface is a rich ecosystem. In the acclaimed Dr. Seuss book *Horton Hears a Who,* an elephant named Horton detects a tiny civilization living on a dandelion. Like the dandelion, our bodies—and every other surface, both animate and inanimate—host tiny communities of beings. Your skin may crawl when you think about miniature life that calls you home, because these microbes are actually crawling (and slithering, spiraling, and sliding) on your skin.

The microbial samples collected from sports fans went on journeys to laboratories near and far. Some of the samples were sent to the Argonne National Lab's Jack Gilbert, who is painstakingly constructing an earth microbiome map. No one yet knows where to find hot spots of microbial biodiversity. There are more plants and animals in the tropical equatorial areas of earth. Are there similar geographic patterns to microbial diversity? Or, could we be creating microbial Amazon rainforests in stadiums, schoolyards, malls, and other places bustling with well-traveled people?

Some of the samples from sports fans, as well as swabs of national landmarks like the Liberty Bell, went on a longer journey. For the first leg of their trip they were cultured in the lab of Jonathan Eisen, a professor at UC–Davis. Eisen's lab set the microbes in a gel of nutrients in preparation for the second leg of their journey: their launch in a payload to the International Space Station (ISS) on April 18, 2014. People who looked up at the right time saw the ISS flying by at seventeen thousand miles per hour; inside were seven astronauts taking notes on the growth speeds of microbes in what they called the playoffs of the micro–Super Bowl. The playfields were petri dishes;[66] the

---

[66] Microorganisms are cultivated in petri dishes, invented by none other than Julius Petri in 1887. The originals were glass, and modern ones are plastic, circular dishes with circular lids. Petri dishes have a thin layer of agar-based growth medium. (Agar is a substance extracted from red algae seaweed, *Rhodophyta.*) The agar base was developed at the suggestion of Fanny Hesse, the wife of an assistant (Walter Hesse) working alongside Petri; she used agar to make fruit jellies, and she knew it would not melt under the incubation temperatures needed to grow bacteria.

referees were the astronauts. NASA supported the project because it might be useful to know how bacteria grow in zero gravity in preparation for eventual plans to send astronauts on longer trips to places like Mars. The playoffs revealed that only one bacteria grew better in space and only one type grew worse. By and large, gravity (or lack thereof) didn't seem to matter for something so small. So far it seems the high levels of radiation in space don't matter either—at least not for growth rates.

Astronauts also sampled high-touch surfaces on the ISS and sent those to the Eisen lab, where they use DNA sequencing to identify the types of microbes that have colonized the space station. The ISS is a great place to take samples and make comparisons to buildings on earth, because the radiation levels are high in space and the source populations are limited to spacecraft and the humans and food on them.

Cavalier and Eisen, both out-of-the-box thinkers, concocted this out-in-space plan. Mark Severance, the director of the Human Space Flight Network at NASA's Goddard Space Flight Center (who also volunteers as the Science Cheerleaders' "Space Guy"), christened the project with the mother of all acronyms, Project MERCCURI (Microbial Ecology Research Combining Citizen and University Researchers on the International Space Station). The organization that brought the Science Cheerleaders together with Gilbert and Eisen to create Project MERCCURI was SciStarter, also founded by Cavalier. SciStarter is an online site that matches, free of charge, tens of thousands of curious people with citizen science projects. The SciStarter menu boasts thousands of projects and filters to help individuals find projects that align with their interests, skills, location, and hobbies.[67] (You are now halfway through this book. If you are not yet involved in a citizen science project, visit SciStarter.com and find the project of your dreams. Better yet, create a SciStarter account and easily participate in many projects of your dreams.)

---

[67] In the interests of full disclosure, I should note that Cavalier and I have a grant from the National Science Foundation to build SciStarter2.0. If you want to engage in citizen science, join the SciStarter2.0 community!

In previous chapters, we met folks drawn to citizen science through games, or matched to projects by their hobbies devoted to birds, butterflies, ladybugs, weather, and stars. The origin of citizen science in microbiology is different. There was never a large preexisting pool of microbe enthusiasts who stroll down paths with microscopes around their necks or petri dishes in their pockets. There haven't been microbe enthusiasts calling for the long-term monitoring of microbial population trends, as with birds and butterflies. For most of human existence, microbes were unknown to us. The only way to see microbes with the naked eye is when they grow in large colonies, like mold on bread. When people spot mold, they are more likely to toss it into the garbage than grab a field guide to identify which type it is.

Citizen science in microbiology began when technology got cheap enough and fast enough in sorting and identifying this previously invisible world that the scale of research could handle the high volume of samples crowds can collect. Citizen science with the invisible world features short-term projects driven by specific research questions and leading relatively quickly to answers to mysteries, such as how many types of microbes inhabit our skin. Microbiologists are casting wide nets into unexplored invisible frontiers. Eventually they will hone in on promising areas. In the meantime, microbes in space collected by sports fans with the help of cheerleaders may not qualify, believe it or not, as hands down the most bizarre activity in this arena. There may not be rivals in other disciplines, but in microbiology, stranger things can happen.

Holly Menninger, director of public science at North Carolina State University, and Rob Dunn, an applied biology professor at NC State, operated a series of short-term projects housed under the umbrella Your Wild Life. These projects were rapid experiments, each inspired by a pressing question, followed through by data collection and analysis, and then completed.

In 2011, with the launch of the Belly Button Biodiversity Project, the intrepid participants who donated samples by swabbing what looked like a giant sterile Q-tip into their navels contributed to the discovery of hundreds of new bacteria. Dunn and his colleagues found over 2,300 types of bacteria in their first sample of donations from just sixty navels. They learned that when we contemplate our navels, most of us are likely ruminating on about seventy different types of bacteria. A group of eight microbial types, referred to as oligarchs, are predictably common in the navels of many people and abundant when present. The remaining thousands of types were each quite rare, often native to a single navel. The microbes tucked into our navels are a representative sample of the microbes covering all of our skin, and we appear to be island archipelagos hosting many hidden biodiversity treasures.

Dunn's research group hoped patterns of microbe diversity in navels could be explained by age, sex, ethnicity, "innies" versus "outies," and the frequency of belly washing, but none of these hypotheses panned out. Other than the common oligarchs, Dunn could not predict what microbes were likely to be found in a given umbilicus.

Dunn went back to participants for more data and uncovered that where you grew up slightly influenced your navel fauna. The baguette-eating French host a composition of navel bacteria that are different from those of bagel-eating New Yorkers and again different from biscuit-eating North Carolinians. Yet there was still a lot of variation left to explain. When they felt completely stumped, Dunn turned out the data to the public with one big shrug. "Have at it," he said.

Sharon Berwick, a mathematician at the University of Maryland, was accustomed to finding patterns in big data, and decided to take a look. She is preparing a research paper with her findings. Belly buttons sort out according to whether they are dominated by bacteria that metabolize in the presence of oxygen (aerobic) or the absence of it (anaerobic).

The belly button biodiversity research informed the next iteration of questions, such as how our skin microbes may affect our susceptibility to mosquitos. The sweat we produce does not have an odor. Rather, our body odor comes from our individual microbiome—could the odors of microscopic critters living on us attract the bigger, buzzing critters that suck our blood? In hope of answering this question, people have been donating their skin microbes to science.

This navel biodiversity research also branched into the study of armpit microbe communities and whether the suitability of armpit habitat is shaped by deodorants and antiperspirants. Led by Julie Horvath at the North Carolina Museum of Natural Sciences, the study asked volunteers who normally use hygiene products to stop for at least a couple of days and those who normally don't use these products to use them daily. The deodorants and antiperspirants achieved their intended outcomes: when someone started using the products, the richness of their microbial armpit community declined. Similarly, when deodorant and antiperspirant use was stopped, bacteria density increased to become similar to those who normally don't use these products. Notably, the types of bacteria living in the armpits of typical product users who abstained briefly were different from those who in the long-term go au naturel. Those who normally do not use products have armpits dominated by *Corynebacterium*, while the armpits of those who briefly stopped using the products were quickly overrun with *Staphylococcaceae*.

Unfortunately, inferences about what these patterns mean are limited because researchers are only identifying bacteria to these basic types rather than to the details of the precise species present. To put this in perspective, identifying an animal as *Panthera* tells me it isn't a house cat (*Felis catus*), but doesn't tell me if it is it a tiger (*Panthera tigris*), lion (*Panthera leo*), leopard (*Panthera pardus*), or jaguar (*Panthera onca*). You might be wise to treat all *Panthera* with the same caution, but that's not necessarily the case with microbes. Some types in the Genus *Staphylococcus* are beneficial, while others (*Staphylococcus aureus*, for example) are sturdy pathogens that rampage through hospitals. Similarly, most *Corynebacterium* are harmless, but one type causes diphtheria.

As is common in science, the findings of one study raise more questions than they answer. Dunn's philosophy, he explains, is that "the next question should be motivated by our findings, not by a desire for new data. Until a new question is asked, there's not much value in new data." That's a practical philosophy because, with microbes and their DNA, it is easy to quickly experience a data deluge.

The longest running project within Your Wild Life was Wild Life of Our Homes, which was designed to identify which factors determine whether a home is a teeming metropolis of diverse, invisible life or a desolate boondock. Menninger estimates that after two years of sampling, they probably have enough data for ten years of analysis to make sense of it all.

To participate, people signed up via the Internet and received microbial swab kits in the mail. Each kit included a protocol and sterile swabs to collect samples from designated areas around their home. In homes, the designated areas were initially inner and outer doorframes, kitchen counters, and pillowcases. From the pilot study, Dunn and Menninger found that different surfaces in homes equate to different habitats. A food habitat includes kitchen and eating spaces. A human body habitat is associated with where people spend time, like beds and bathrooms. An outdoor habitat, such as window sills and doorframes, is related to locations where life enters from the outside. After the pilot study, more designated areas were added, including toilet seats and TV screens.

What governs the distribution of microbial life? To collectively find out, when each participant sealed swabs and mailed them to North Carolina State University, they also filled out an online questionnaire to provide details about their home and its occupants. When the swabs arrived in Raleigh, Menninger placed them in a freezer to keep the bacteria from growing. Once they had a critical mass to make the processing cost-effective, they started to tally the diversity of microbes in the samples based on DNA sequences.

The logistics of genetic processing of microbial samples are challenging, because microbes are everywhere, including on researchers and the lab bench. Microbiologists take great care so that local microbes don't contaminate samples of microbes taken from afar. For the Wild Life of Our Homes project, the lab of Noah Frierer, a collaborator at the

University of Colorado–Boulder, was entrusted with the process. Frierer, in turn, used a combination of technicians and a robot to meticulously turn samples collected by citizen scientists into data on the genetic code of microbes.

Dunn and Menninger found no two countertops contained the exact same microbial diversity. Ditto for pillow cases, doorframes, and toilet seats. Though not identical, species composition was more similar within each habitat (a pillowcase in one household compared to a pillowcase in another household) than within each house (a pillowcase compared to a kitchen counter).

Dunn's group examined many factors that failed to predict which homes would be populated like microbial cities and which would be like a microbial ghost town: not the number of people living in the house, nor the presence of children, pesticides, or carpet.

Man's best friend was the answer. Dunn's group compared the microbiome of homes with pet dogs to the microbiome of homes without shaggy inhabitants. The presence or absence of a dog in the house explained almost half the variation in the types of bacteria on pillows and TV screens (microbes get stuck with static cling). Also, the composition of fungi in homes was influenced by regional fungal communities outside. Where you live mattered for fungi, but not for bacteria. Who you live with matters for bacteria. For one variable, dogs, to explain half of the variation in an ecological pattern is as much a contribution as one of your parents made to explain your own personal traits. Dunn speculates that dogs reconnect us to the diversity of species that occur in the great outdoors but which don't usually make it into our homes. They may even help our immune system.

The Wild Life of Our Homes project may one day revolutionize microbiome research, but so far it has been taking incremental steps through microterrain. Menninger explains, "There's still a lot of growing pains and newness in the field. Things are really evolving and moving quickly. The field is still moving through the descriptive phase of the research, such as who's there and why. I think the big stuff will come next when we think about applications to health and forensics."

With regard to forensics, the Wild Life of Our Homes project could operate like a Sherlock Holmes clue. Microbe communities are unique enough to function like old-fashioned fingerprints. The Homes (Holmes!) project was able to predict the geographic location of a sample based on its fungal community to within two hundred kilometers. Further sampling with indoor microbes could reduce the size of the fingerprint, making it useful for solving crimes. A murderer may wipe his fingerprints from the weapon or victim, but he can't remove his signature composition of microbes. Remember, we each have a distinctive body odor and breath, and these are autographs of unique microbial communities.

With regard to health, Menninger is referring to the hygiene hypothesis, an idea that cleanliness may not actually reside next to godliness. Better hygiene has contributed to decreases in the rates of infectious disease, but an excess of these hygienic practices may be contributing to soaring incidences of allergies and autoimmune disorders (e.g., type 1 diabetes). In the last decade, asthma has reached epidemic levels in industrialized countries. Similarly, atopic dermatitis, also called eczema, has increased in the last three decades, affecting 15 to 30 percent of children. For decades, an industry waging war on microbes has grown, selling antibacterial soap, disinfectants, and hand sanitizers. Now the low microbe diversity in our overly sterile, antibacterial crazed lifestyles may be backfiring on immune systems. Although plenty of types of microbes are harmful, microbiologists have identified plenty more that are helpful or neutral to our health. For the majority of invisible life, they don't know which are good and which are bad. It's uncertain whether we benefit from exposures early in life that challenge our immune system, or whether exposure causes lasting harm.

Microbiologists involved in citizen science anticipated facing a challenge in convincing people that most microbes were not villains and were worthy of their attention. To the contrary, people have taken the hygiene hypothesis to heart, and then some. Participants commonly have a desire to find miracle microbes on their bodies. Unexpectedly, researchers have to conscientiously dispel misconceptions about the sensational powers of microbes. They have to repeatedly ask participants to not interpret the personal results of their microbiome composition as medical advice.

Eisen operates a website dedicated to dispelling myths about new industry-driven claims and misinformation about microbes as medicine. The field is teetering on the edge of a solid understanding of the microbiome. Answers to many microbe mysteries are tantalizingly close, "but not close enough," warns Menninger. To date there is only one valid microbe-based medical treatment, which is fecal transplant to treat those infected by *Clostridium difficile*, a microbe found in the gut. As it sounds, the procedure involves transplanting feces from a healthy donor to the colon of an unhealthy patient. It works because the healthy gut microbes prevent *C. diff.* from taking over the intestines of the patient. When valid treatments involve poop donations, how can the average person differentiate a far-fetched claim from a substantiated claim? "Misinformation from businesses wanting to make a buck play on the incredibly personal meaning of each citizen science sample. That is dangerous," Menninger cautions. "Participants tend to see their results as diagnostic, and they are not. They are only explanatory in the collective." Although it is an appealing notion, one can't simply analyze the microbes on his or her pillowcase or toilet seat and draw conclusions about personal health.

By and large, current studies of the microbiome mimic the 1800s-style natural history of explorers Meriwether Lewis and William Clark: observing, classifying, and piecing together observations into bigger patterns. Researchers and citizen scientists journey out, collecting and identifying specimens from as many unexplored places as possible to map the diversity of (miniature) life in the world. With microbes, every place is unexplored terrain, from the New York City subway (currently being mapped with PathoMap by Weill Cornell Medicine) to air and soil (by Noah Frierer's lab at the University of Colorado) to individual armpits (with Your Wild Life at North Carolina State University and the North Carolina Museum of Natural Sciences) to the International Space Station (by UC-Davis). Now that we have the techniques to uncover invisible worlds, citizen science and traditional science are in the same boat, overwhelmed by the microbes that exist everywhere. The field is focused on basic biogeography, the study of places and what lives in those places. The next phase will be to find out why microbes live where they do and to study the consequences of their distribution—particularly as regards human health.

In microbiology a unifying concept has been germ theory, the notion that diseases are caused by microbial organisms. The idea that entities too tiny to see were causing diseases was first proposed in 1546 by Italian scholar Girolamo Fracastoro in what he described as spores that could transmit disease. He was an atomist—that is, he believed that chemicals were made of smaller components that we could not see, called atoms. So it wasn't a big leap for him to say the same for disease agents. Before germ theory, people assumed that diseases spread by miasma, the term for noxious air from decaying matter. While breathing in a gust of odor from roadkill may induce a sudden visceral sickness, miasma was incorrectly fingered as the culprit that brought epidemics of cholera and the Black Death.

Unlike germs, germ theory spread slowly. For example, in the mid-1800s Austrian obstetrician Ignaz Semmelweis noticed that new mothers were dying of fever after giving birth with the help of doctors and medical students, but not after births assisted by midwives. When he convinced doctors to wash their hands with chlorinated lime water before entering the birthing room, which was on the rotation *following* the autopsy room, death from childbirth went down from 18 percent to 2 percent at Vienna General Hospital. Still, the medical establishments rejected Semmelweis and his ideas about germs. Mules, mountains, and scientific establishments rank in the top three on the list of things that are nearly impossible to move. Remember this—centuries of stubbornness—as we soon begin to consider how citizen science might reorganize the scientific establishment today.

But first, let's look at how microbes were finally seen.

It was in the 1670s—one hundred years after Fracastoro, and almost two hundred years before Semmelweis—that microorganisms were first seen by the human eye. Although Fracastoro had speculated the existence of disease-carrying spores too small to be seen, when microorganisms were finally seen, the discoverer studied them with fascination and awe, never associating them with germs.

The honor of the achievement of seeing microorganisms for the first time goes to Antonie van Leeuwenhoek (credited as the father of microbiology) in 1676. I want to highlight Van Leeuwenhoek's career because he straddled the blurry line between what we'd consider an amateur scientist today and a man of science of his day. He was not part of the scientific establishment; only after landing a job as a civil servant at age forty did Van Leeuwenhoek come into contact with intellectual elites like Robert Hooke,[68] who was part of the nucleus that formed the Royal Society of London. While men of science conversed in Latin, Van Leeuwenhoek only spoke, read, and wrote in modern Dutch. Nevertheless, he published over five hundred papers, keeping at it until he was more than ninety years old and on his deathbed. He was an anomaly among natural philosophers because he never taught students or visited a university.

Van Leeuwenhoek differed from important scientists of his day in another pivotal way. Hooke, for example, was interested in using microscopes to understand more about already familiar organisms like lice, silkworm eggs, and the wings of flies. Leeuwenhoek, like citizen microbiologists today, sought out otherwise invisible life. Instead of searching navels and doorframes like Your Wild Life, he searched milk, blood, and water in ponds, lakes, and streams for miniature life never before encountered.[69] His blue-collar background gave him an advantage that he carried close to his chest: he had developed a method with

---

[68] Some attribute the honor of seeing the first microorganism to Hooke, for his viewing of a microscopic fungi: white mold on leather, which he reported at a Royal Society meeting in 1663 and again in his well-regarded 1665 book *Micrographia*.

[69] He was the first to describe the sperm cells of animals, including humans, and he discovered the fertilization process, that sperm enters the egg cell.

which to make powerful lenses. In his younger years as a cloth merchant, he used lenses to judge the density and weave of fabric, and he must have learned a thing or two from his hands-on practice.

Van Leeuwenhoek's secret technique involved putting a flame to the middle of a small rod of soda lime glass until it pulled apart like taffy into two long whiskers of glass. Then he would apply the flame to one of the whiskers until it formed a very high-quality glass sphere. The smaller the sphere, the greater its ability to magnify. With sophistication he would grind and polish the spheres to perfection, some becoming as thin as one millimeter and with a radius of curvature of 0.75 millimeters. He would mount a lens in a hole in a brass plate, in line of sight of a sharp point to hold the specimen. The focus was adjustable with a couple of screws. His contraptions were three to four inches long, with lenses that magnified as much as two hundred times. The top-of-the-line compound microscope of the day, which used two lenses, was invented in the 1590s and was available to Hooke but only magnified twenty or thirty times.

Being secretive was common then and, some might argue, now. For example, at the time, not only were scientific communications expected to be in Latin but it was also customary for men of science like Hooke to stake claim of their emerging discoveries through anagrams, short unreadable messages with the words or letters out of order. They announced their breakthroughs as if in a jumble puzzle in the Sunday paper in order to prevent anybody else from claiming they made the discovery first. They would print an anagram for anyone to read, and later the author would expose its meaning. For example, in 1655, Christiaan Huygens reported his discovery of a moon around Saturn (now called Titan) by telescope with an anagram about the moon's orbital period: "ADMOUERE OCULIS DISTANTIA SIDERA NOSTRIS UUUUUU CCCRRHNBQX." In 1656 he unscrambled his anagram, leading to the Latin sentence *Saturno luna sua circunducitur diebus sexdecim horis quatuor*, which translates to "A moon revolves around Saturn in sixteen days and four hours." He had another for rings of Saturn, which he revealed only after three years because that is how long it took him to secretly validate his finding: "encircled by a ring, thin and flat, nowhere touching, inclined to the ecliptic." Aptly, Christiaan Huygens's own name was an anagram: "Anarchy, this genius."

Galileo sometimes used anagrams too, but he didn't in his 1631 book *Dialogue Concerning the Two Chief World Systems.* A year later he was under house arrest and required to renounce the Copernican view of the solar system in which the planets orbit the sun instead of orbiting earth. It is no small wonder that scientists mostly shared discoveries only with each other at conferences and then policed themselves in journals, creating traditions that separated scientists from outside scrutiny, but also from laypeople.

Van Leeuwenhoek's powerful, secret lenses, coupled with his outsider status, put him in the same predicament as Horton the elephant in Dr. Seuss's book. In the 1670s, Van Leeuwenhoek began reporting on the existence of single-celled organisms, which he called animalcules. With the help of a translator, in one paper he submitted to the Royal Society (this was almost a century before William Whewell received his Royal Medal, as I noted in the introduction) he reported animalcules on pepper grains he had soaked in water for three weeks. Members of the Royal Society were skeptical; after some kerfuffle, they decided to settle the matter by sending a team of respected observers to visit and confirm Van Leeuwenhoek's observations. The situation is not unlike what citizen scientists face today when they need to provide photographic documentation of their observations (for a discussion of this, see chapter 8). The members of the society needed only a glimpse of Van Leeuwenhoek's secret lenses, and his discovery of animalcules and many other tiny discoveries, to recognize his competency. After a decade of discovery he was finally chosen to be a jolly good fellow of the Royal Society in 1680.[70] Later,

---

[70] Van Leeuwenhoek's acceptance into the intellectual elite turned the budding scientific enterprise on its head, if only temporarily. He also turned another patriarchal system on its head—that of honey bees. Men of science at the time assumed that each hive was governed by a king bee, but Van Leeuwenhoek used his lenses for micro-

in the 1680s, he explored where even Your Wild Life doesn't venture and found animalcules in the white matter stuck to his teeth. Though Van Leeuwenhoek is considered the father of microbiology,[71] the field did not bear any immediate children. It was many years, perhaps a century, later that microbes would again be the subject of intense research interest, this time as sinister sources of disease.

The twenty-first century's academic children of microbiology are rebelling against their father Van Leeuwenhoek by abandoning secrecy and adopting open science practices. They have no choice, given the data deluge and citizen science participants who are impatient for answers about what invisible life inhabits their belly buttons, armpits, and windowsills.

According to "The Tao of Open Science for Ecology," a paper coauthored by Stephanie Hampton of the Center for Environmental Research, Education, and Outreach in Pullman, Washington and her colleagues, open science is the "concept of transparency at all stages of the research process, coupled with free and open access to data, code, and papers." Researchers who adopt open science practices view themselves as stewards, rather than owners, of data. Dunn displayed open science when he shared the entire data set on navel microbes. Open science practices go hand in hand with citizen science because participants expect to be kept in the loop—and rightly so.

For some scientists like Dan MacLean, adopting open science practices with colleagues was a gateway to citizen science and being open with the public. MacLean, the head of bioinformatics at the Sainsbury Laboratory and the John Innes Center, was drawn to open science because it is the quickest way to battle a fungus that causes disease in trees in the United

---

dissection and pulled numerous immature eggs from the honey bee ruler, thereby dethroning the king and proving to the scientific community that hives are led by queens.
[71] Van Leeuwenhoek also helped overturn the idea of spontaneous generation, the notion that smaller forms of life arise from nonliving things. He observed reproduction in the colonial flagellate *Volvox*, an algae, signifying that even animals of the tiniest proportions become parents.

Kingdom. The disease agent is *Charlar fraxinea* (the name of the anamorph stage),[72] or Frax, which is causing a rapid loss of ash trees (called ash dieback) from treasured English woodlands.

MacLean began an open research program to sequence the genome of Frax in hopes that the genetic code would reveal solutions to cure or prevent ash dieback. With a quick solution as the primary concern, he decided that the research strategy needed to be ultracollaborative. Usually scientists will generate data, analyze that data, and collaborate with a few trusted coworkers; when they have made sufficient use of the data in a series of publications, then they *might* share the data with other scientists. MacLean describes the traditional system bottleneck, in which work is vetted by a small number of people and then available to a restricted readership via journal subscriptions, as "filter, then publish."

MacLean describes open science as "inverting" the old system; he prefers "publish, then filter" because that allows faster progress with a larger community of minds. Through the open science paradigm there is no need to wait to share data and insights; it is better to share an incomplete story with a community of researchers who are ready and waiting for the release of data. Immediate review and criticism is music to a scientist's ears because it translates into rapid advances.

To figure out what makes the fungus so deadly to ash trees, MacLean's lab collected DNA from infected trees, sequenced it, and released the data to crowdsource the analysis. Through a project called Open Ash Dieback, MacLean continually releases genomic data on the fungus to a public website hub as soon as his lab sequences it, before he and his colleagues even begin any analyses. He says this research approach is "in the vanguard of a new movement."

---

[72] Fungi are classified into species based on their sexual structures. A fungus can occur in a telemorph stage, which is the sexual reproduction morph, like fruit; or it can occur in the anamorph stage, which is the asexual reproduction morph, like mold; or the fungus can be a holomorph, both at once, like the whole mushroom. For many years a fungus could have two names, one for the anamorph stage and another for the telemorph. The dual naming system was discontinued in January 2013 by the International Code of Botanical Nomenclature when it adopted the guiding principle "one fungus, one name."

Rapid DNA sequencing and similar genetic methods produce too much information for scientists to handle, and even too much for scientists from multiple disciplines tackling the problem together. The doors of open science need to be propped ajar so the public can stroll in. Thus, MacLean's next step was to use citizen science to engage more minds in data analysis. MacLean's group developed a Facebook game called Fraxinus, similar to Phylo (see chapter 6), through which players can, without any training, figure out solutions to multiple sequence alignment problems. A distributed crowd of friends is like a human supercomputer. In the first year, players spent over nine hundred days on Fraxinus. For the trickiest areas of the Frax genome, people proved better at finding alignments than did computer algorithms.

Open science works. The first open dieback data led to the certain identification of the pathogen *Hymenoscyphus pseudoalbidus*. Within two months, researchers had built two libraries of genomic DNA of *H. pseudoalbidus*, annotated for searchers by chemists who specialize in fungicidal compounds. Chemists soon created a candidate target for fungicides. Specialists in fungal secondary metabolic pathways used the genomics data to identify fungicidal compounds that were potentially toxic to Frax. Other analyses suggested the disease is entering the tree by decaying its wood. Another finding quickly identified two mating types in one small area, which meant that the pathogen was from multiple origins. These types were then found in tree nurseries and recent plantings, suggesting that the sale of seed and saplings was one cause of the spread of the disease. Open science—open sesame! Like magic, tactics to slow the spread of the malady were quickly in sight.

Open science does just what the phrase implies: opens the discovery process by adopting a style that breaks traditions. Public science, open science, and citizen science are already creating new traditions of sharing, and these are all departures from the old ways. The fundamental steps of the scientific method haven't changed much, but a short trip through history can illustrate the slow but steady progress in opening the curtains that keep new knowledge hidden from public view. Let's take a look at past restrictions on the spread of knowledge in order to prepare for a future of the highly accessible, open, and public generation of new knowledge.

It is disappointing to have a paper full of one's scientific revelations rejected from a journal, and negative reviews from one's peers can cut to the quick. Yet it beats house arrest, prison, or being burned at the stake. During the Renaissance, the closest validation system to modern-day peer review was limits placed by state and church censorship. A tragic example is Michael Servetus, who, in 1553, was the first person to correctly describe human pulmonary circulation. Unfortunately, he reported the details of heart ventricles (blood passed from the right to the left through the lungs) in *Christianismi Restitutio* (The restoration of Christianity), a book in which he also expressed his religious views, which were considered heretical at the time. He rejected the idea of a trinity in favor of a unified God, and also the principles of infant baptism. He wasn't throwing the baby out with the bathwater, but simply rejecting the concept of predestination, insisting that individuals start out good and only through their words or deeds can become unworthy and thereby condemned by God. John Calvin and the Geneva Council convicted Servetus of heresy, for which he was burned at the stake with what was believed to be the last copy of his book chained to his leg.

Originally, most of the copies of his book were unbound and tied in bails to look like paper. The ruse didn't work, and almost all of the copies were destroyed as soon as they appeared for sale. Fortunately, three copies of his book somehow escaped destruction and appeared decades later when they were more easily distributed. Servetus's studies of human physiology propelled medicine, and his religious views became a key part of the Unitarian Church.

The Royal Society founded a journal called *Philosophical Transactions* in 1662 and submissions were reviewed by the editor (and, in the exceptional case of Van Leeuwenhoek, a group of trusted visitors). Editors assessed whether contributions followed standard practices of creating scientific knowledge, such as eliminating possible explanations based on observations until the last remaining explanation could be accepted as the most likely—a principle first articulated in 1620 by Francis Bacon in his *Novum Organum Scientiarum* (New instrument of science). Continual improvements to typewriters throughout the 1800s, which led to carbon copiers in 1890 and eventually Xerox copiers in 1959, also led to slow changes in journal practices. The ability to easily make multiple copies of an article for review meant that every manuscript submitted to a journal could be scrutinized by several other scientists for independent opinions on the research. So-called peer review became common practice for all academic journals in the 1960s. Many great innovations, including theories propagated by Charles Darwin and Albert Einstein, were not peer reviewed in the manner that is the norm today.

Eurocentric science has grown into a monolithic and colossal field. As gates went up around the system of creating knowledge, it also went up around the knowledge itself. Print journals required subscriptions; when journals went to online formats, most retained a subscription fee. With the Internet otherwise full of free information, having the products of taxpayer-funded research located behind a paywall has raised some fundamental questions about intellectual property and freedoms. The argument isn't only about what's fair. Closed science creates roadblocks that make it slower than open science. And new knowledge is not spontaneously generated, but built on the foundation of what's already known, on the shoulders of giants—unless the giants, or the legacy of their works, are hidden behind paywalls.

Aaron Swartz (who gave us an alternative analysis of Wikipedia users in chapter 2) was a founder and developer of Reddit in his teen years; he grew into an adult advocate for open knowledge.[73] In January

---

[73] As a teen he was also integral to the creation of the Creative Commons, founded in 2001 as a system of free copyright licensing. With a Creative Commons license, creators give permission to share their works on conditions of their own choosing. Most simply want attribution when their works are used or built upon. The goal is to keep the Internet free, creative, and open.

2013, at the age of twenty six, he committed suicide, unable to cope with harsh prosecution by the criminal justice system for committing a victimless crime: Swartz had downloaded and made public millions of academic articles from an archive called JSTOR. Although Swartz had library access to all the articles in JSTOR, sharing these articles with others violated the archive's terms of service agreement. During a two-year ordeal, Swartz was prosecuted under the Computer Frauds and Abuse Act, among other regulations. He was fighting thirteen felony counts and facing up to thirty-five years in prison and a $1 million fine. Swartz was an idealist committed to Internet freedom; his actions were political, not for profit. At the time, no one imagined he would get charged with a federal crime, much less that he'd face prosecution with no option for a plea bargain. In fact, he was so confident that he would not meet serious consequences for his actions that he did not try to hide his activism: he placed the computer that was accessing JSTOR in a place that obviously had surveillance cameras. Everyone wised up to the seriousness later. When I speak on the phone with John Wilbanks, a proponent of open science, he reflects, "the copyright lobby was never going to let anyone get away with what Aaron did."

Swartz's act of civil disobedience, and the case against him, added momentum to an open-access campaign called Access2Research. About a year before Swartz's suicide, the quartet of Wilbanks, Heather Joseph (executive director of the Scholarly Publishing and Academic Resource Coalition), Michael Carroll (a law professor at American University), and Mike Rossner (then director of Rockefeller University Press), launched Access2Research to convince the White House to mandate that journal articles arising from taxpayer-funded research be freely accessible on the Internet. Their strategy involved flexing their right to petition the US government, which is granted by the First Amendment of the Constitution (in the same clause as the right to assembly, free press, free speech, and freedom of religion). In 2011 the White House developed a procedure for petitions and criteria for government response: the petitions platform is called We the

Author Caren Cooper, visible to the public in her lab behind a glass wall at the North Carolina Museum of Natural Sciences, is curating eggs of invasive House Sparrows. The eggs were collected and sent to her by volunteer birdwatchers as part of a citizen science project with dual purpose: to evaluate methods for minimizing damage to protected birds by House Sparrows; and to assess the potential for House Sparrow eggs to function as a biomonitoring tool allowing communities to quickly index levels of environmental contaminants. (*Photo courtesy of NC State University*)

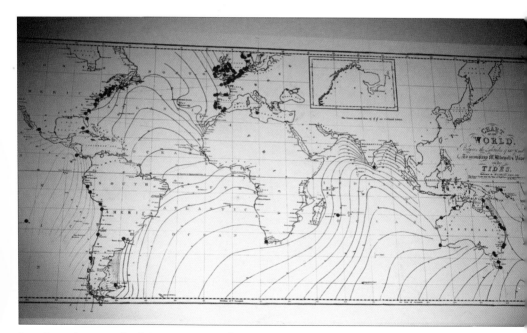

In a pioneering example of citizen science, William Whewell orchestrated the Great Tide Experiment, a data collection event that largely contributed to his award-winning research on ocean tides. In the Great Tide Experiment, thousands of ordinary people simultaneously monitored the tides with tide gauges at hundreds of sites across the world, providing Whewell with millions of observations with which he could craft and test his theories. (*Photo courtesy of Michael Reidy/The Crowd & The Cloud*)

Retiree Dan Matthews, a dedicated volunteer in Moncton, New Brunswick, Canada, contributes rain and snow measurements to CoCoRaHS, the Community Collaborative Rain, Hail, and Snow Network, every day, even in the coldest depths of a Canadian winter. CoCoRaHS (discussed in chapter 1) is a volunteer network for precipitation data collection, which allows for more precise meteorological forecasting. *(Photo credit: Art Howard/The Crowd & The Cloud)*

CoCoRaHS founder Nolan Doesken proudly shows off the project's rain gauge installed in the White House garden, symbolizing the Administration's support.

(Photo courtesy of CoCoRaHS)

Angel and Mariel Abreu are knowledgeable and enthusiastic birders, running a guiding company with the appropriate name Nature is Awesome. They accompanied *The Crowd & The Cloud* host/presenter Waleed Abdalati (center) during filming in the Florida Everglades. Birders participate in citizen science projects such as NestWatch, a nest-monitoring program in North America that was founded in response to the catastrophic decline in bird species due to the widespread use of the pesticide DDT. NestWatch and other ornithological citizen science projects are discussed in chapter 2.
(*Photo credit: Geoff Haines-Stiles*/The Crowd & The Cloud)

Scott Eustis (left), Shannon Dosemagen, and Jeff Warren documented the BP oil spill in the Gulf of Mexico as part of the Public Lab response to the Deepwater Horizon disaster, using open source software to stitch together imagery collected with low-cost cameras flown below kites and balloons. Since the BP oil spill, Public Lab (discussed in chapter 2) has expanded across the United States and internationally, capturing high-resolution imagery of environmental pollution.
(*Photo credit: Nathan Dappen*/The Crowd & The Cloud)

Using the Nature's Notebook app from the National Phenology Network, teens from Rockaway, Queens, New York, participate in habitat restoration efforts in the nearby Gateway National Wildlife Refuge after Superstorm Sandy, demonstrating how STEM learning can be applied in the real world. Here they are carefully plotting where pollinators can be found on storm-resistant plants; pollinator habitats and conservation is further discussed in chapter 3.

*(Photo credit: Sean Feuer/*The Crowd & The Cloud)

In Europe, Tiger mosquitoes spread diseases like West Nile, malaria, and dengue fever, as detailed in chapter 3. In the city of Barcelona, the "Mosquito Alert" app encourages citizens to report where mosquitoes may be breeding by submitting photos with GPS coordinates. Initially developed to attack the Tiger mosquito, the app is being updated to help combat Zika and other mosquito-borne diseases. Lessons learned from its development in Spain are being applied to a new NSF-funded "GLOBE Observer" MOSQUITO WATCH citizen science app for the United States.

*(Photo credit: Geoff Haines-Stiles/*The Crowd & The Cloud)

Andy Westphal is adapting the "Stardust@home" user interface—the online citizen science project discussed in chapter 4 that allowed volunteers to identify interstellar dust tracks for NASA—to support "EyesOnALZ," a project that's working to speed up the search for a cure for Alzheimer's.

*(Photo credit: Sean Feuer/*The Crowd & The Cloud)

Clark, Wyoming, resident Deb Thomas (right) and one of her neighbors became part of a citizen science project using "Bucket Brigade" techniques to sample the air close to local fracking sites, which the neighbors suspected were emitting unhealthy and dangerous chemicals. Her data was included in a "Warning Signs" study from community activists, and in a peer-reviewed article in the respected journal *Environmental Health Letters*. "Bucket Brigade" trains citizen scientists to use relatively simple sensors and a rigorous protocol to collect air quality samples close to industrial sites, which are then analyzed by professional laboratories. As mentioned in chapter 7, water quality deteriorates in the vicinity of fracking. *(Photo credit: Nathan Dappen/*The Crowd & The Cloud)

Trout Unlimited volunteers in North Central Pennsylvania monitor local streams and sample water quality and other stream measurements using an EPA-developed checklist to collect baseline data and alert local governments to spills. As mentioned in chapter 7, over 100,000 people in the United States monitor local water quality. (*Photo credit: Andy Quinn*/The Crowd & The Cloud)

Glaciologist Ulyana Horodyskyj stands with members of the Sherpa-Scientist Initiative (SSI) in the Nepalese foothills of the Himalayas after rebuilding a weather station damaged during the 2015 earthquake. SSI trains the Sherpa to monitor the weather station instruments year-round, contributing to science and helping to understand flooding in their local environment. The SSI functions in a manner similar to that of the Silalirijiit Project, described in chapter 7, which compares climate data from weather stations with local Inuit observations in Clyde River, Nunavut, Canada.

(*Photo credit: Benjamin Pothier*/The Crowd & The Cloud)

Sonam Sherpa works with science technologist Patrick Rowe to learn research techniques involving robotics and innovative sensors as part of glaciologist Ulyana Horodyskyj's Sherpa-Scientist Initiative (SSI).　　　　　　　　　　　　　(*Photo credit: Benjamin Pothier*/The Crowd & The Cloud)

Philly Unleaded is a citizen-led effort aimed at testing the lead levels in the drinking water of Philadelphia residents. After the drinking water crisis in Flint, Michigan, (discussed in chapter 9) Philadelphia resident Jonathan King was concerned about lead levels in his city's water. He reached out to Marc Edwards, the Virginia Tech civil engineer who had assisted citizens in Flint by testing their polluted water. Jonathan found other concerned neighbors via a Facebook group, and the Philly Unleaded citizen science project began.　　　　　　　(*Photo credit: Sean Feuer*/The Crowd & The Cloud)

Just as North Carolina citizen scientists fought back against air pollution in chapter 9, Cassandra Martin from Oakland, California, combated air pollution in her community. She joined with neighbors to count trucks emitting pollutants as part of the West Oakland Environmental Indicators Project, collecting data with a low-cost air pollution sensor, as depicted above. After presenting the community's citizen science data to the City and Port of Oakland, the routes of trucks emitting pollutants were changed to lessen impacts on homes, schools, and day care centers.

*(Photo credit: Andy Quinn/*The Crowd & The Cloud)

People. In those early days, using the site to demonstrate sufficient public support for a cause, and getting the president's attention, meant getting twenty-five thousand signatures in a month (now that the site has caught on, the threshold for the president's attention is 100,000 signatures).

In late May 2012, when Wilbanks, Joseph, Carroll, and Rossner created a White House petition with We the People, the movement went viral on social media. In the petition's first day they acquired an unprecedented eight thousand signatures. They reached their goal in two weeks and acquired over sixty-five thousand signatures before the campaign closed at the end of one month. Internal White House negotiations progressed, although they paused during election season, resuming at the beginning of 2013, which coincided with Swartz's suicide. Forty days after Swartz's death the White House and its Office of Science and Technology Policy announced the new executive order: all US federal agencies that have significant budgets for research and development must develop open access policies within twelve months.

Joseph called the directive "a watershed moment." In a news article, she said, "The directive will accelerate scientific discovery, improve education, and empower entrepreneurs to translate research into commercial ventures and jobs. It's good for our nation, our economy, and our future."

In the wake of Swartz's suicide, the scientific profession is slowly reshuffling economics and meritocracy systems in such a way as to promote open access to the products of science. With so much politicking, prosecution, and sacrifice going into the fight to make the *knowledge products* of scientific research freely available, it is time to remind ourselves that citizen science goes one step further: people gain access to the *process of making of new knowledge.*

So far we've mostly encountered examples of citizen science where participants collect observations or solve puzzles as prescribed by scientists. These are referred to as contributory projects. Echoing former president John F. Kennedy, participants don't ask what science can do for them, but what they can do for science. By following the scientists' lead, contributory citizen science tends to reinforce the existing authority of scientists, but with important changes in the status quo operations of the scientific establishment, adding transparency and openness. In the following chapters we'll see citizen scientists doing even more: setting research agendas, devising questions, and making decisions based on their interpretation of the results. In this way citizen scientists tend to shake up the existing structure of the scientific establishment by demanding to be shareholders in the science market. We'll see that the same way citizen science gives people access to explore the mystery of our connection to tiny invisible life is the same way that it is supporting an enormous revolution through which access to science helps balance disparities of wealth, education, and power.

We start the transformation of science here, where microbiome research is disarming and open enough to engage citizen scientists. If a scientist feels all right with the absence of arrogance, embracing the silliness of studying invisible organisms in bellybuttons and cheering sports fans to gather samples, then people respond with a generous attitude, let down their guard, and enter the door that science is holding open.

Those nostalgic for the eras of discovery and exploration now navigate frontiers of unexplored terrain on microscopic levels. To deal with floods of data, scientists need to see themselves as part of a network in which many brains are better than one. Wilbanks humbly confides, "To state it badly, the paradigm that needs destruction is the idea that we as scientists exist as unnetworked individuals . . . We need to network ourselves and our knowledge. Nothing else we have designed to date as humans has proven to scale as fast as an open network."

Scientific answers are part of the solutions to humanity's biggest and most complex threats, like climate change. Indigenous knowledge also can play a role. How communities use citizen science to address problems provides lessons that could make or break our future. As we'll see next, whether living on thinning ice in the Arctic or dwelling in corporate finance in penthouse offices, real solutions are found when people become stewards of the earth.

CHAPTER 7

Conservation Biology
Taking Stock

You are not Atlas carrying the world on your shoulder. It is
good to remember that the planet is carrying you.
—VANDANA SHIVA

GLOBAL CLIMATE CHANGE PRESENTS ONE OF THE BIGGEST CHALLENGES TO
the quality of humanity's future. In the arctic, the Inuit have firsthand experience with climate change impinging on their daily lives:
they are literally on thin ice. Some Inuit collaborate with scientists to
carry out research to better understand climate change and share their
wealth of knowledge about climate. These are valuable contributions
that challenge common notions of who can make new knowledge because the Inuit do not have their own written language.

Most of the world's population lives in temperate and tropic environments where a typical weather forecast includes the high and low
temperature and odds of precipitation: "Today you can expect a high
of seventy, a low of fifty, and an eighty percent chance of rain." From
those estimates we make our daily personal decisions, such as what we
wear and whether we might jump in puddles today. Those who live in
arctic climates are not so temperature and precipitation oriented. It's
almost always freezing, and there can be snow in the air at any time,
either falling from the sky or blowing up from the ground. For people
at extreme high latitudes, factors in daily personal decisions can be,
quite literally, blowing in the wind.

Wind is one of the weather conditions that matters most and is less
predictable with climate change. As Elaiya Mike, an Inuk in Iqaluit, the
largest city in the Canadian territory Nunavut, reports, "Nowadays we

are getting wind from everywhere. The winds are shifty and constantly changing their point of origin. The weather signs point toward a clear, calm day, but the wind suddenly whips up, and that is how it seems to operate in this day and age."

Inuit hunters in Nanuvut travel on land, ice, and open water to harvest ringed seal, caribou, Arctic char, narwhal, polar bear, and ptarmigan, to name a few. All told they hunt about twenty-six species of mammal, bird, fish, and invertebrates. In places like Clyde River (known by the native Inuit as Kangiqtugaapik), a community of about 850 people, the hunters deliver a share of their harvest to the elders and invite others in the community to come and collect some.

Since 2009, Shari Fox Gearheard has led the Silalirijiit Project (Silalirijiit means "those who work with or think about weather"). Gearheard, a geographer and research scientist employed by the University of Colorado–Boulder, has lived in Clyde River since 2004, and has established a collaborative working relationship with many Clyde River hunters and elders. Through the Silalirijiit Project she and a team of expert hunters and scientists compared climate data and Inuit knowledge of wind speed and direction over time. It was not surprising to the team to discover that scientific data and Inuit knowledge do not always align to tell the same story; instead they tell complementary stories.

Inuit have been reporting an increase in the variability in wind speed and direction since the 1990s. Inuit hunters pay close attention to the wind, which affects the amount of blowing snow, the size of the waves, whether fog is likely to roll in, and the speed of moving ice, among other things. Inuit say that the prevailing wind direction used to be from the northwest. A steady prevailing wind creates relatively stable snow drifts (*uqalurait*) that are a reliable aide to navigation. Now, however, the Inuit report that the prevailing wind direction changes frequently. The direction of the snow drifts change, and people can become lost; with navigation thus more dangerous, some Inuit now carry personal locator beacons and satellite phones.

The data from the local weather station reveals no detectable change in the prevailing wind direction and barely any change in wind speeds over time. Unlike expert Inuit hunters, weather instruments are stationary, near the sea on flat terrain at the Clyde River Airport. The wind instrument there records hourly wind speed via a two-minute average taken at the beginning of every hour. Analysis of the data have shown that on an annual basis, the proportion of high winds (those exceeding thirty kilometers per hour, at which point travel is dangerous) had not changed over time, and the proportion of low winds (lower than twenty kilometers per hour, when travel is safe) had decreased slightly.

The weather station is not malfunctioning; nor are the Inuit wrong. They are observing different phenomena. The weather station is not representative of the fjords where the hunters travel. Inuit are sensitive to sudden and unpredictable changes in wind because they affect their safety and navigation. Indigenous knowledge and scientific knowledge are different, and Gearheard stresses that neither is superior; both need to be interpreted and applied correctly.

The lack of alignment between scientific and local knowledge in this example highlights the need for citizen science in rapidly changing environments (though neither Gearheard nor the Inuit who collaborate with her call what they do citizen science). An important lesson from Gearheard's study is that there are valid forms of knowledge, such as Inuit knowledge, apart from conventional scientific research. Bringing together scientific information and indigenous knowledge can be extremely useful and illuminating, but what exactly is combined and how this is done needs to be carefully considered. As Gearheard explains to me, there is a lot more to *saying* that bringing traditional knowledge and scientific knowledge together is important (many studies *say* this), because actually *doing* this work can be challenging.

Many researchers over the years have interviewed Inuit and found recurring stories, particularly with regard to another important weather variable: sea ice. For many of us, ice is a cube of frozen water that we take from the freezer and put in a tall glass of lemonade. For people in the far north, ice is the landscape. Sea ice varies in strength, thickness, salinity, and texture. Across the landscape, ice is a patchwork of types.

New ice is typically smooth and thin (up to thirty centimeters), and ice in its first year is thicker (up to two hundred centimeters), and often rough; older ice is the thickest (over two meters) and has lower salinity and greater strength.

Many Inuit have noticed that sea ice has become thinner and is present for shorter amounts of time than before. Alooloo Kautaq, an Inuk hunter in Clyde River, describes how snow and seawater now mix into *punnirujuk*, which is similar in consistency to lard and deteriorates more easily than iced used to, thus posing a danger. The Inuit have also noticed that permanent snow patches, *aniuvat*, are smaller. The seasons have changed in length and timing. Simeonie Amagoalik, an Inuk from Resolute Bay, provides an example of indigenous knowledge: "I used to go egg hunting, but now it is too dangerous to travel by ice, so I cannot go to the places that I used to go to. I think it is mainly the ice on the sea that has affected me the most." Inuit know the sea ice is thinning based on how it affects their daily lives and livelihood, rather than by measuring its actual thickness. Jaypeetee Qarpik from Pangnirtung recalls a youth spent traveling by ice as early as November, if not sooner, but now can boat through open water during the twelve days of Christmas. What's relevant to the Inuit is not the numeric thickness of the ice but how it affects their hunting; researchers, on the other hand, need the exact measurements. Gearheard wanted studies of ice to be relevant to both Inuit and scientific communities.

For five years, starting in 2006, Gearhead led the Siku-Inuit-Hila (Sea Ice–People–Weather) Project. Part of the project involved Inuit collecting measurements of sea ice thickness and sharing the data with Gearheard and an interdisciplinary team of university- and community-based researchers. Once a week during the sea ice season, hunters traveled to measure the sea ice conditions at designated monitoring stations.

Each station was an eight-meter square with four corners made of stainless steel cable suspended from wooden stakes and frozen through the ice and into the seawater below. In the middle were nine wooden stakes in the snow, which a volunteer would examine from outside the grid in order to measure snow depth without disturbing the snow cover. After recording snow depth, they heated each corner cable by applying AC voltage across the top of two cables at a time. Under the ice, the seawater conducted the electric current, which heated both cables. Once they yanked the hot cables free of the ice, the Inuit would haul them up until a wide weight attached to the cable would bump the bottom side of the ice. They would measure the length of the newly exposed cable, from which they computed the thickness of the ice upon which they were standing (and upon which their way of way of life depends).

We may not stand directly upon sea ice as the Inuit do, but all our lives, anywhere on the planet, do rest upon sea ice. The reason scientists take a particularly strong interest in Arctic sea ice is because it influences all life on the planet. Sea ice covers about 15 percent of the oceans for part of the year, but that relatively small fraction has an enormous impact on global climate. Sea ice is frozen ocean, unlike icebergs, glaciers, ice sheets, and ice shelves, which are freshwater and originate on land. Sea ice reflects solar radiation (heat) back into space, so areas of the ocean stay cold. Sea ice is the opposite of paved blacktop surfaces, which absorb heat and feel hot. The ocean works as a conveyor-belt style circulation system that essentially moves weather across the planet. That conveyer belt is driven by sea ice: new sea ice is salty, but as it ages, it pushes salt into the ocean underneath and the saltier, colder water under old sea ice is dense and sinks. It moves along the ocean floor toward the equator, pushing the mid-depth warmer waters from the equator toward the poles. The warm air above the tropical ocean waters moves right along with those waters, making midlatitudes comfortable for us and our crops.

The data collected by the Inuit at the monitoring stations catalog some sea ice basics—time of freeze-up, time of breakup, sea ice thickness, and snow thickness—and are informative to local residents. Plus, with the data collected at sea ice stations, the research team can figure out the processes governing growth and melt at the surface and at the bottom of the sea ice.

Of equal value to the research team is the knowledge the Inuit elders and hunters have from oral history and personal daily experience on the sea ice. For example, the sea ice study identified different drivers of ice growth and melt at different locations. The town of Qaanaaq proved to be the most unexpected; it is the coldest town in the region, but its ice was 20 percent thinner due to an unexpectedly high rate of bottom melting. It was ascertained that a warm ocean current must be preventing ice growth, making the region more susceptible to climate change than places farther south. This information helps the town plan adaptation. The entire town depends on thinning sea ice so that hunters can access the water for hunting, travel, and to harvest icebergs for fresh water in winter when the local creeks dry up. If the ice gets too thin, the town won't be able to harvest passing icebergs with the front-end loader trucks that bring the bergs to the water-processing facilities. They are starting to plan now for the inevitable time when the ice becomes too thin.

Collaborative efforts between scientists and community members has not always been the norm in the Arctic but, increasingly, visiting scientists are recognizing Inuit knowledge, and Inuit are taking more leadership roles in research projects and programs. For generations, Inuit hunters and elders have used traditional weather forecasting methods based on a variety of indicators including cloud patterns, wind, animal behavior, and the twinkling of stars. Climate change means unpredictable weather patterns; if the system is unpredictable or more variable, does it make it more difficult for elders to pass on their knowledge (called Inuit Qaujimajatuqangit, or IQ) to younger generations?

Where indigenous people remain, effective conservation strategies involve preserving and studying indigenous knowledge, as well as helping the people find alternatives when rapid environmental change causes long-standing traditions to wane. The Bushmen of the Kalahari of the Desert use an app called CyberTracker to record how they track animals while hunting. The Bushmen are not like deer hunters in the United States, who often hook a chair ladder to the side of tree and sit and wait for deer to come within range of their gun. Bushmen are long-distance runners, and they work as a group to separate an animal from its herd and run it in the desert heat while keeping themselves in the shade when possible; the animal dies of heat exhaustion, but the hunters recover.

Louis Liebenberg, an innovator from South Africa, developed the CyberTracker app in 1996 and has been upgrading it ever since. Trackers don't read words, but they read animal tracks in the sand and other signs in nature, so it wasn't a stretch for them to read computer symbols or icons. Liebenberg calls tracking an art and hypothesizes that it uses the same reasoning skills as scientific methods, suggesting that tracking provides insights into the origin of science. Many researchers have posited that the human brain is a paradox of evolution by natural selection. If our ancestors' brains were adapted to deal with the daily problem solving of hunter-gatherers, why our brains deal with math and physics? As biologist and naturalist E. O. Wilson puts it, "That is the great mystery of human evolution: how to account for calculus and Mozart." Monitoring animal signs is important for understanding the art of tracking and the potential origin of science, preserving indigenous knowledge, and the conservation of wildlife.

Sometimes the vulnerability of indigenous communities is not from pervasive climate change but from logging by multinational companies, which in turn opens up forests to wildlife poachers and allows urban-based businesses to dominate local trade. Fortunately, some industries, like the Congolaise Industrielle des Bois (CIB) want to meet standards set by the Forest Stewardship Council, a program that certifies responsible forest management. As part of their certification, the CIB aims to minimize the negative impacts of logging on indigenous people. The Bayaka Pygmies in the Congo live as hunter-gatherers in forest tracts the size of small US states; they are stigmatized and marginalized by their urban neighbors. Even though they don't subscribe to principles of landownership like the majority of cultures in the world, they have many resources in the forest that they depend upon for their livelihood. To protect these resources, including sacred medicinal trees and freshwater springs, they turn to citizen science. The Bayaka Pygmies use Global Positioning System (GPS) devices, with the icons and images displayed similar to CyberTracker, to geotag and map natural resources important to them. CIB uses the maps created by the Pygmies to guide their selective logging and the timing of cutting schedules. The Pygmies also use GPS units to report instances of poaching and the discovery of illegal roads. In this way, citizen science gives voice to the Pygmy people in a country where they are otherwise marginalized and excluded.

Across the globe there are people acting as stewards of natural resources. Stewardship can involve a range of ordinary activities, from placing a bird bath in a backyard to harvesting fish only over a certain size. When it comes to conservation, citizen science is no longer about hobbies or games pursued in one's free time. It is a necessity, and can be a critical part of how one choses to spend time.

Finn Danielsen and his colleagues are based in Copenhagen, where they organize participatory monitoring for natural resource management with numerous communities—particularly those that neighbor protected areas such as national parks and marine reserves. In 2010 they published a paper quantitatively comparing the effectiveness of participatory monitoring relative to monitoring carried out solely by teams of professionals. In an e-mail exchange, Danielsen explains, "We had encountered skepticism from a number of people who did not believe that community members can monitor biodiversity." He recalls a specific incident in which visiting biologists completely discounted participatory monitoring during conversations with government staff. Fortunately, the government staff recognized that the academic biologists lacked knowledge of the realities of managing protected areas. Even though it was unfounded, the criticisms by academics drove home the message to Danielsen and his colleagues that they needed to show a quantitative comparison, published in a highly esteemed journal, in order to convince academic natural scientists of the value of participatory monitoring.

What they found was that participatory monitoring helps local people make decisions, like whether or not to go fishing in certain areas, to pass a village bylaw banning the hunting of wild pigs during the breeding season, or to only permit the harvest of shellfish that are old enough to reproduce. The data from professional efforts have informed larger institutional changes, like the ratification of the Kyoto Protocol to reduce carbon dioxide emissions, changing international agreements on fisheries and subsidy policies, and placing threatened species on the International Union for the Conservation of Nature's Red List of Threatened Species. Professional monitoring influences policies on large scales, while participatory monitoring influences real decision making at home.

Danielsen also found that the time between data collection and action differs between professional and participatory monitoring. When natural resource management involves community members in environmental monitoring, people act on the results quickly— usually within months. When only professionals are involved it can take years, even decades before findings are used to inform any type of decisions.

Danielsen explains, "It is not professional scientists who are slow, but it is the professional scientists' *process* which is slow." He says that scientists are too far upstream in the decision-making process. The process is slow even when their studies quickly produce technical reports for government agencies or nongovernmental organizations (NGOs) rather than slowly produced peer-reviewed publications. He admits, "The professionals' research agenda is often out of touch with local community needs."

Scientists simply preparing a report doesn't create a compelling case. Participatory monitoring functions like a forum that inspires discussion and thought, followed by rapid action. "When community members are involved, the route to decision making looks very different. It is much shorter," explains Danielsen. "When community members are involved, just by being involved, observing changes and discussing their findings together with each other, they immediately start thinking about possible actions that need to be taken. Participatory monitoring provides a forum for interpreting trends, for identifying solutions and for taking action. This is very important."

For example, Danielsen notices that communities will often be drawn into discussions about setting quotas on how many individual animals can be taken from a wild population. When community members are themselves directly involved in collecting and interpreting data on animal populations and proposing the size of the quotas, they are more likely to cooperate with the authorities on overseeing that the quota restrictions are followed in practice. They are more likely to call government staff if they see someone violating the regulations.

Participatory monitoring is a key part of successful conservation strategies for resolving conflicts between humans and desirable wildlife like snow leopards in Asia and koalas in Australia.

Snow leopards are an endangered species native to the Himalayas and protected by international treaties. The Nepalese government took action to reduce the hunting of snow leopards and their main prey, blue sheep. As a result, snow leopard populations in many areas have rebounded but, unfortunately, more snow leopards leads to an increase in human-leopard conflict. Local communities feel that they must bear the brunt of government conservation through losses in their livestock.

Snow leopards live at an elevation of between three thousand and five thousand meters, and are indicators of the health of the mountain ecosystem. In this range, families eke out a living by growing crops and herding livestock, neither of which is easy in an area with cold, harsh weather, and rocky and steep terrain. Herders free-range native yaks, cattle, sheep, goats, horses, mules, and donkeys; these animal provide dairy, protein, fiber, and leather; dung supplement for fuel; dung for fertilizer; transportation of goods and people (tourists); and labor as draft animals for the tilling of land. When snow leopards kill livestock, the economic loss can amount to as much as a quarter of a family's annual income. Herders sometimes slay snow leopards in retaliation for their monetary shortfall. The healthiest snow leopard populations were observed only around Buddhist monasteries, which was evidence that retaliation was a problem; observant Buddhists don't kill animals, so the monasteries functioned as holy nature preserves. Conservation biologists realized the government efforts to recover this endangered species would ultimately fail without a way to address the needs of the local people.

A rebounding snow leopard population taking some livestock is a common problem repeated in all the countries where snow leopards live: Afghanistan, Bhutan, China, Nepal, and Pakistan. The most effective solution to human–snow leopard conflict in each of these countries has been community insurance, which compensates for livestock loss. Similar programs exist in the United States, such as NGOs that create a wolf compensation trust for ranchers in the northern Rocky Mountains. Over a couple of decades, conservation organizations paid out over $1 million to ranchers to compensate for wolves taking livestock before the 2009 Omnibus Public Land Management Act transitioned the program to be run and funded by states and tribes.

In central Asia some insurance schemes to compensate for snow leopard loss rely on tourism funds. A model system in Pakistan involves a collective insurance fund and an ecotourism fund that cofinances the insurance compensation.[74] In Nepal, there is a model that is 100 percent community owned and managed, which buffers it from relying on external funds other than seed grants to establish initial credit. In India, communities added a twist of payment for ecosystem services: herders agree to keep their livestock off some pastures for the grazing of blue sheep and other leopard prey; the pastures set aside for wildlife function like preserves, and local villages receive payment at fair market value for grazing land used by the wildlife.

---

[74] Pakistan was on the so-called hippie trail (aka the overland trail) in the 1970s. It still gets about one million tourists per year now, despite continued conflict.

In some locations, communities have added a citizen science component to their insurance scheme. For example, in Ghunsa, Nepal, herders use motion-triggered cameras to monitor snow leopards and blue sheep. Called camera trapping, the activity provides important insights into the ecosystem. With the help of herders, scientists have learned that snow leopards keep blue sheep populations low, which actually makes more grazing available for local livestock. The cameras also provided data on how many snow leopards are in a given area. Perhaps most important for long-term conservation, as villagers learned more about the snow leopards through pictures of these elusive animals, the snow leopards went from being considered pests to being a source of pride; the camera trapping functioned like a snow leopard publicity team.

In neighboring Bhutan, yak livestock are more at risk of disease than snow leopard predation, but similar camera trap programs are growing. Tshewang Wangchuk is the executive director of the Bhutan Foundation and a PhD graduate student studying snow leopards. He hopes that citizen science will make the relationship between herders and snow leopards more meaningful. With just a little training and equipment, the herders can set up and maintain cameras while out herding their yaks. They don't expect anything in return, and they get excited by the awesomeness of the pictures.

Wangchuk tells me about one twenty-five-year-old herder in Bhutan who got amazing pictures of three snow leopards feeding on one his yaks. When the herder reported the data, he wasn't complaining or interested in retaliation; he was thrilled! "Not one, not two, but three snow leopards!" he boasted. It was undoubtedly a hardship for him to lose a yak, but that hardship was overshadowed by joy in seeing the lost yak supply food for a big family of snow leopards. In many countries and cultures, we undervalue joy. In Bhutan, joy is a highly prized commodity and the country quantifies and reports its wealth not simply in terms of gross national product but also with an index of gross national happiness.

With the highest diversity of felines on earth, Bhutan wants to be a premier tourist destination. One of the most popular hiking trails, the Jomolhari Trek, travels through prime habitat for snow leopards and blue sheep. Focusing on the snow leopard hot spots, Wangchuk selected two communities along the trek to participate in snow leopard monitoring with camera traps. Soe Yutoed has twenty-eight households and Soe Yaksa has eighteen; the villagers are yak herders, and the area is mostly above the tree line. Tourism helps villagers benefit from snow leopard conservation; as Wanghuck sees it, "Tourism and citizen science transform snow leopards from a liability to an asset."

Koalas are another species of tourism interest that can cause conflict when overly abundant. On a single Wednesday in November 2012, over 450 people in South Australia reported on the locations of over thirteen hundred koalas in a citizen science project called the Great Koala Count. Researchers at the University of Adelaide used the data to model koala occurrence and identify suitable habitat. The places where koalas thrive include Kangaroo Island, the Mount Lofty Ranges, and the tips of three peninsulas of South Australia. But this wasn't always the case.

When Europeans settled in South Australia in the 1830s, koalas were present and doing fine in the southeastern part of the state. A market developed for their thick pelts, and by the 1930s millions had been killed for the fur trade, and koalas were extinct from South Australia. As their numbers got low, starting in the 1920s, koalas were reintroduced to Kangaroo Island and other parts of South Australia. Now the reintroduced koala populations have rebounded with a vengeance, and there are an estimated 200,000. In the Mount Lofty Ranges alone there could be over 100,000 koalas. The Great Koala Count helps to estimate their abundance and distribution on the mainland, and the data support research that can inform management options. The conservation is ultimately determined by what management options people support. For example, by 2001 the Kangaroo Island population was an estimated twenty-seven thousand. Koalas have become a suburban species; it might be difficult for someone outside of Australia to imagine it, but koalas are often considered pests. The government management plan included culling in some states, like Victoria, but when wildlife managers proposed culling the population on Kangaroo Island, mainlanders voiced strong opposition.

Wildlife managers and mainlanders compromised in a program to sterilize koalas. Kangaroo Island is now site of one of the largest wildlife sterilization programs in the world. Meanwhile, in New South Wales, koala populations are declining from rampant spread of chlamydia, which can manifest as a sexually transmitted bacterial disease that can cause blindness and infertility. In some areas, more than half the koalas have chlamydia, and wildlife managers are again proposing culling to stem the spread of the disease in hopes of restoring a healthy population. Similar to management of suburban white-tailed deer in the United States, culling for any purposes is often not only unpopular but met with strong public resistance. Time and again, conservation requires not only scientific insights on what works but public support on what's acceptable.

A study led by Bianca Hollow and her colleagues at the University of South Australia and at the Department of Environment in South Australia, examined how the Great Koala Count influenced the role of public support on the development of policy for koala management. Hollow compared three groups: participants in the Great Kola Count; those who had heard about and registered with the Great Koala Count but didn't participate (referred to as onlookers),[75] and those who had not heard of the Great Koala Count (considered to represent the general public).

The groups were asked for their opinions toward various ways of managing koalas.

Participants and onlookers differed from the general public in a few ways; for example, they favored banding trees, which involves placing a strip of tin or aluminum around a tree trunk in order to prevent

[75] Twelve hundred people registered, and 472 participated; compared to the general public, participants tended to be older, well-educated women.

koalas from climbing it and defoliating it. The general public seemed to favor doing nothing, or perhaps translocations, but definitely not banding trees. Similarly, when people ranked koala management priorities, the general public ranked reducing car collisions with koalas as a priority, though participants ranked raising awareness and research into disease as priorities. When the survey focused on cars and the management options of providing road ladders, promoting safe driving, adding more road signs, or minimizing koala access to roads, the general public favored minimizing access, while participants favored the addition of road ladders. For all responses, onlookers tended to be like participants or fall in between participants and the general public.

The views of participants (and onlookers) were not representative of the general public. Hollow contends that the citizen scientists represent highly informed members of the public and therefore are as valuable, or even more so, for informing policy than the general public. Many participants reported that they learned about koalas and koala management, and even changed their opinions about management, during the Great Koala Count. Over 65 percent of participants and over 60 percent of onlookers reported that they had learned something new from their involvement in the Great Koala Count. Furthermore, about 19 percent of participants and almost 38 percent of onlookers said that the Great Koala Count caused their opinions to change, most often related to koala management and particular options like fertility control. Policy makers could engage citizen scientists in dialogues to inform decisions.

The phenomenon of elevated civic engagement among citizen scientists extends beyond koalas, notably with volunteer water quality monitors. There are over 100,000 people in the United States who monitor local water quality, and hundreds of thousands worldwide. They are not monitoring at the spigot, as in the Flint Water Study (see chapter 9), but at the source: in rivers, streams, and lakes. Volunteer water monitoring has a long history similar to the projects featured in part 1 of this book. The primary difference being that projects in part 1 involved national data sets created via a central location for all local data to be shared, whereas volunteer water data has typically been managed at the state level, collected locally, and largely used locally.

Ted Ludwig is a volunteer water monitor in Colfax, Wisconsin. He also recruits and trains volunteers, instructs schoolchildren, and helps scouts earn badges through water monitoring programs. Ludwig is a retired marine and a retired postal worker; at seventy-three years of age, he is brimming with energy and is outside every day. For example, when he and his wife take a trip to visit her parents, who live 230 miles away, they don't hop in a car. They load up their panniers with food and his bike trailer with camping gear and pedal their bicycles for three or four days.[76] There are many streams and lakes for which volunteers can monitor water quality. Not surprisingly, Ludwig does them all.

At the basic level, Ludwig uses a net to sample macroinvertebrates in the water. This is his favorite task with kids because they are often shocked that bugs live in water. By identifying the types, he can report a Benthic Index, a calculation based on the combination and amounts of different types of aquatic bugs. Some types of macroinvertebrates are tolerant of low-quality waters, and others are intolerant; which type is present reveals a lot about the health of the aquatic system. Ludwig runs a test for dissolved oxygen, and estimates turbidity with a special tube. To measure stream flow, he uses the float method, measuring the width and depth of a stretch of a stream and recording how long it takes a ball to float a known distance on that stretch. At the next level, he'll install a thermistor to continuously measure temperature from May to October, which is something he says helps scientists determine whether a stream should be classified as a cold water or warm water one and indicates what species of fish it can harbor. At the highest level, Ludwig will do special projects like sampling phos-

---

[76] Since 2011, Ludwig has been participating in a citizen science project in which he mounts an acoustic monitor to his handle bars and bikes at a steady pace for fifteen miles on the Red Cedar Trail after dark, recording bat calls.

phorus and road salt runoff. He enters all the stream data into the database of the Wisconsin Department of Natural Resources, called the Surface Water Integrated Monitoring (SWIM) system. Anybody in Wisconsin making water decisions uses SWIM data. Professionals and volunteers alike add data to SWIM.

Kris Stepenuck, an assistant professor at the University of Vermont, studies water quality monitoring. As part of her dissertation research in Wisconsin, she surveyed over three hundred coordinators in the United States to find out how their programs have influenced natural resource policy and management. A paper by Stepenuck and Christine Overdevest reported that new and experienced volunteers did not differ in their knowledge of water monitoring topics, but they did differ in other significant ways: people with lots of experience in water monitoring were more likely to be politically active, to be part of larger social networks, and to have stronger feelings of community connectedness than others.

One common policy outcome was that projects identified illegal bacterial discharges. Also common was that projects were able to upgrade the protection status on a monitored stream by identifying its risks. Sometimes program data informs designation of National Wild and Scenic River status, or it might be used to gain gear restrictions for creek fishing. Ludwig has helped get streams designated as "impaired" so that they receive attention from programs that encourage farmers and other landowners to adopt practices that limit stream pollution and sediment runoff. Ludwig keeps monitoring, but he has not yet seen a stream successfully restored. He muses, "Humans are the only critters who try to wipe out their own living areas."

Wisconsin supports its volunteer monitors because it relies on them for baseline monitoring. The state provides small grants for equipment, and it provides training. Other states are not as supportive; Pennsylvania had rigorous volunteer monitoring programs in the past, but cut funding in 2009, diminishing the resources available to the more than eleven thousand volunteer monitors in the state. Around the same time, from 2007 to 2012, the state issued over ten thousand permits for hydraulic fracturing.[77] Coincidence? How much policy makers, regulators, and managers want an engaged public varies from place to place.

New knowledge is the main product of citizen science. Civically engaged people, stewards of natural resources, communities empowered with new knowledge, and social ties: these are a few of the by-products of citizen science that are key to conservation and environmental justice (as we'll see in chapter 9).

Some scientists are concerned that the desire for the by-products of citizen science, and their social impact, will compromise scientific integrity. The journal *Nature* published an editorial called "Rise of the Citizen Scientist" in 2015; though the article praises many aspects of citizen science, it expresses concern about the potential for conflicts of interest: "One reason that some citizen scientists volunteer is to advance their political objectives. Opponents of fracking, for example, might help to track possible pollution because they want to gather evidence of harmful effects." As we've seen, people want to find pollution *if* pollution is present, which is very different from people simply wanting to find pollution.

The editorial went on to allude to the Koala Count in another potentially problematic example: "When Australian scientists asked people who had volunteered to monitor koala populations how the animals should be managed, they found that the citizen scientists had strong views on protection that did not reflect the broader public opinion."

That's not evidence of conflict of interest that leads to bias, that's evidence of success of citizen science in creating an informed public—at least among a subset of the public most vested in the issue. The success of democracy rests with an informed public. Funding agencies expect good research to produce these types of broader impacts, and that includes by-products of citizen science that support social change.

---

[77] Hydraulic fracturing, also known as hydrofracturing or fracking, is a controversial process through which large volumes of water and chemicals are injected into the earth under pressure to fracture shale that may contain natural gas or oil. It is commonly understood to be a process that can ruin water quality in surrounding areas.

The concern about biased research and loss of objectivity is not just an issue affecting the credibility of citizen science but something that conservation biologists have always faced. Conservation biology is a value-laden discipline, based on a commitment to conserving biodiversity. Conservation biologists struggle with how objective research and advocacy can coexist. George Wilhere, a biologist with the US Fish and Wildlife Service, has argued that the answer is to avoid inadvertent advocacy, which he defines as "the act of unintentionally expressing personal policy preferences or ethical judgments in a way that is nearly indistinguishable from scientific judgments." As we've seen with koalas, decision makers arrive at a verdict by combining objective scientific findings with societal values. How that combination happens is key to maintaining trust in science, and now in citizen science. The typical equation would have scientists adding knowledge and the public adding values; decision makers would then mix these to get the optimal policy outcome. With citizen science, people add both knowledge and values; to complicate matters, their pursuit of knowledge appears to alter their opinions and possibly even their values. Citizen scientists need to avoid inadvertent advocacy that arises from conflating knowledge and opinions; this is the same tightrope walked by conservation biologists.

New scientific information is essential in providing humanity with solutions to our problems, but it alone is not sufficient. Just as necessary for producing workable solutions is a way to mix social values and scientific knowledge, both of which are constantly evolving. With citizen science, participants can help find scientific discoveries that feed into a science-policy nexus. Equally important is that they can also bring their perspectives, ideas, values, and opinions into a values-policy nexus. If experiences in citizen science alter what goes into the values-policy nexus, that's probably a good thing for a well-informed democracy.

Ultimately conservation must involve people because we can't tell birds when to fly south or snow leopards where to eat. We can only change our own behaviors based on our understanding of our individual and collective influence. We are veritable forces of nature. People are at the center of conservation not just because we create the problems, or even because we are capable of studying the problems, but because only we can create solutions. Citizen science gives us the insight and foresight to manage natural resources sustainably.

Given that participation in citizen science, and perhaps even being an onlooker, has the power to change peoples' views on conservation priorities and policies, why not foster those connections with people driving the types of economic development that often lead to disruptive environmental change? Given how citizen science can change opinions of crowds, could it change the view of climate change deniers in pivotal positions? If only stockbrokers were willing to note the wind speed and direction, or lawyers could note when ice forms and thaws. Too bad bankers don't collect data about trees!

Or do they?

In the corporate district of London, I passed through security and rode the elevator to the fortieth floor of the HSBC bank building. I was directed into a conference room where I stood at the window, gaping at the London skyline. I was waiting to meet Bill Thomas, who was pioneering the use of citizen science to change the hearts and minds of high-level banking executives.

At this point I believed that citizen scientists could solve global problems, but I had not thought deeply about the professions held by citizen scientists. I knew citizen scientists could be found among teachers, the clergy, community organizers, plumbers, dental hygienists, and used car salespeople. I'd written about young citizen scientists, senior citizen scientists, and incarcerated citizen scientists. One commonality among participants was that their experiences with citizen science took place at or near home. To learn about people who do citizen science far from home, I had contacted Earthwatch, which arranges for people to spend their vacations as field assistants on research projects in remote, beautiful places around the world; part of the Earthwatch fee goes into the budget of the research project. To my surprise, the people at Earthwatch introduced me to their corporate program, which tackles climate change.

"Hey there," Thomas said as he walked in. I was disappointed that he was American, because I'd traveled so far. He seemed to appreciate an American visitor, which gave him a respite from tea breaks.

Thomas created the Sustainability Leadership Programme for bank executives. In partnership with Earthwatch, the program sends executives to be citizen scientists at designated Earthwatch field sites around the world.

Thomas was direct in summing up why he created citizen science opportunities for corporate executives: "We have to get leaders out of the building and into the field. In conference rooms, they wouldn't remember a damn thing. When their hands get dirty, when they are contributing to meaningful research—that puts them in different frame of mind so they are open to changes."

I thought Thomas was speaking figuratively, but then he showed me a photo montage of well-manicured hands covered in dirt and displayed for the camera; the hands, he explained, were those of CEOs. He thought back to where the photos were taken, at a reforesting project in Brazil run by Earthwatch scientists: "It rained all week. We were all soaking wet. Ahhh, they loved every minute of it!" The CEOs' lives were changed by getting their hands dirty.

Thomas projected a slide with a quote attributed to Benjamin Franklin: "Tell me and I forget. Teach me and I remember. Involve me and I learn." He explained that the first sentence could describe a typical employee training program and the last sentence could describe why citizen science is effective at changing people.

Anyone who has sat through an employee training program will likely agree that they are not transformative. With the Earthwatch partnership, the exit surveys from HSBC employees in citizen science programs were overwhelmingly affirmative: Yes, the program was of value. Yes, I would recommend it. Yes, I would do it again. HSBC (or anyone else, I bet) has never before encountered a company training program with that remarkable level of success.

Thomas explained that employees tend to dislike traditional team-building exercises, which can feel contrived. Contributing to real science is an authentic, meaningful experience. "A nature walk is not going to do it. They aren't interested in playing in the woods or the water. They want to contribute. They are hardheaded bankers," he emphasized. I don't have a face for poker, which is probably why he continued, with quiet patience, upon detecting my skepticism (or cynicism): "Bank executives are human, like everyone else. They honestly want to make a positive difference in this world."

Before Thomas was a sustainability leader, he was an information technology specialist with HSBC. And he was a climate skeptic. He believed that climate change was happening, but he didn't believe that humans were causing it. He went on one of HSBC's corporate programs to the Smithsonian's Environmental Research Center in Maryland. There he met scientists who explained climate change in detail and answered his questions. As soon as he was able to discuss the topic directly with scientists, he recognized that he'd been accepting a distorted reality as presented by the media sources he chose. Thomas came to realize not only the reality of humans causing climate change but that it is an environmental, business, economic, societal, educational, and research issue.

He described himself as a changed man when he returned to work; he now had a burning desire to create change in the business world. He quickly came to think of the program he had completed, called Climate Champions, as training the foot soldiers. General bank employees like him weren't in positions where they could create change within HSBC; they were coming back highly motivated to make a difference, but then quickly becoming frustrated. They were making changes in their personal lives but not their professional lives.

Thomas's enthusiasm for sustainability led him to become the global head of sustainability engagement for HSBC and begin the partnership with Earthwatch; he created the Sustainability Leadership Programme to train the senior managers (the generals). The graduates of the program are expected to engage others at senior levels in an effort to create significant change in business practices. "I'm not the only one," he said. "Lots of hardheaded leaders come back changed." HSBC field trips engage bankers in one of the largest studies of the carbon storage capacity of forests around the world, and they learn about the importance of forests for mitigating climate change. Thomas understood that seeing charts of data doesn't change people. "We have to get to their hearts, not their heads," he said.

Thomas retired shortly after we met, and one of his former employees, Matthew Robinson, stepped up as sustainability engagement head at HSBC. Under Robinson's guidance, HSBC has continued to be creative in its use of citizen science in training bankers to take leadership roles in sustainability. Social scientists have documented the enormous impact of the Sustainability Leadership Programme, and this model is now being adopted by several business schools.

Since 2010, HSBC's Sustainability Leadership Programme has held more than ninety training sessions, graduating more than a thousand senior executives from over fifty countries, each magnifying their impact from their leadership positions. Robinson has designed the program to be what he refers to as "cross-border, cross-functional, and cross-business." Sustainability leaders learn skills for teamwork and cooperation because sustainability requires collaboration with partners and customers. Collaboration is tricky in the business world because it is easily undermined by competition. Robinson emphasizes that "building a sustainable business that truly creates meaningful change will increasingly be thanks to leaders (and the stakeholders who they interact with on a daily basis) developing a deeper sense of purpose and mindfulness."

Early on, HSBC's sustainability leaders delivered efficiencies within the bank itself. For example, several graduates of the training sessions were later in charge of hiring a firm to manage all HSBC facilities. Their training in the Sustainability Leadership Programme motivated and prepared them to insert into the contract an effective combination of incentives and penalties so that their facilities are managed with energy and waste reductions, thus leading to reduced carbon footprints.

Today sustainability leaders are developing new solutions that resonate with HSBC's clients, helping others decrease their carbon footprints in order to collectively transition to a low-carbon economy. For example, another graduate of the Sustainability Leadership Programme took a closer look at the procurement of paper. He led efforts to favor the purchase of sustainably produced paper products, such as those certified by the Forestry Sustainability Council, for HSBC as well as their partners and customers. To reinforce this, HSBC also invites the senior executives of companies it partners with to partake in the Sustainability Leadership Programme. Given the experiential learning of the citizen science experience, HSBC wants its partners on the same page when it comes to sustainability.

The ways the graduates of the program find to bring sustainability practices to their work is diverse. For example, when shopping for a vendor to provide information technology servers, a sustainability leader at HSBC screened applicants to ensure that all equipment was Energy Star compliant. Sustainability leaders examine the global supply chain and work to assure sustainable practices from source through delivery. Another sustainability leader actively encourages video conferencing in order to lower the carbon footprint associated with air travel. The list goes on, and each of these programs is a reminder that actions can make a difference. Experiential learning through citizen science helped bankers figure out how to use their influential positions in ways that matter.

PART 3
A World Where Everybody Counts

OST OF THE STORIES IN THIS BOOK HAVE EXPLORED HOW HOBBYISTS or science fans work with scientists within a setting where the scientists are ultimately in charge of discovery. The assumption is that scientists benefit from collaborating with the public to achieve the primary product of new knowledge, and we've seen hints of how the public benefits from the by-products. Citizen science, however, can disrupt the paradigm, reexamining who is in charge and how the benefits are distributed. Scientific practice is an authoritative system with which to produce trustworthy knowledge, but it doesn't have to be authoritarian. (And it's not the only game in town, as we we've seen with indigenous knowledge). The further people integrate citizen science into the fabric of their daily lives and the realities they face, the more the by-products of citizen science will be increasingly vibrant and useful for global well-being. In this section we'll meet citizen scientists at the helm of the production of new knowledge. We'll see how people flip the citizen science model: instead of people helping scientists make discoveries, we'll see scientists helping people make discoveries.

CHAPTER 8

Marine Biology
Turtles and Nurdles

UNLESS someone like you cares a whole awful lot, nothing is
going to get better. It's not.

—Dr. Seuss

M Y VISIT TO WRIGHTSVILLE BEACH, NORTH CAROLINA, IN JULY 2012
coincides with a heat wave: daily highs exceed one hundred degrees. I know it is customary to enjoy any opportunity to visit the
beach, but my being is wholly irritated by the heat and humidity, the
sand in my scalp, the salty air on my lips and in my eyes, the sun on
my freckles, and the ceaseless pounding of the surf. I cannot fathom
why 44 percent of the world's population lives near the sea.[78] I'm a
mountain girl; I take to the hills. I'd rather wear flannel and whittle a
stick in the shade of a hemlock tree, listening to songbirds and banjos.

On Friday, July 6, I arrive at 5:55 a.m. at Johnnie Mercer's Pier,
where the sand is strewn with spent fireworks. My clothes are already
sticking to me. I have arranged to join Susan Miller at 6:00 for turtle
patrol of zone 3; we are to walk from Mercer's Pier along the beach
for about three quarters of a mile to public access point 32, and then
back again, while looking for signs that any turtles came ashore during the night. Miller is a volunteer for the North Carolina Wildlife
Resources Commission; state agencies are mandated by the Federal
Endangered Species Act to create and implement a plan for the recovery of all populations of threatened and endangered species that make
use of their states. North Carolina has only two permanent turtle bi-

---

[78] According to the *United Nations Atlas of the Oceans*, 44 percent of the population
in 2010 is equivalent to the entire global population in 1950.

ologists and over three hundred miles of nesting beach, much of it along the barrier islands known as the Outer Banks;[79] thus the necessity for the biologists to rely on about seven hundred volunteers like Miller to find, protect, and study the nests of the threatened loggerhead sea turtles.

A woman toting a camera with a telephoto lens approaches me quickly; thinking it must be Miller, I make eye contact. "Do you mind if I stand next to you?" she asks. "I'm trying to photograph the couple under the pier and don't want them to notice me."

"Sure," I say nonchalantly, pretending her request is run-of-the-mill. Under the pier, the man kneels down on one knee, makes an ungodly long speech, eventually rises, and the couple embraces. The diamond on the engagement ring will last forever—just like, I am about to learn, the microplastics swirling next to them in the surf.

Miller arrives with an empty, reusable garbage sack slung over her shoulder, like Santa Claus down on his luck. I had expected her to carry a clipboard, and maybe a measuring tape, but I soon learn that there are no special tools for this work. Finding sea turtle tracks is pretty simple as long as you encounter the tracks—which look like those of a small dune buggy with bald tires—before the tides wash them away. Miller explains that if we were to encounter turtle tracks, which are not very common on Wrightsville Beach, we would contact the local head volunteer to figure out whether and where there might be a nest.

Miller hands me a smaller trash bag. "Let's get started," she chirps. After explaining how to look for the tracks of sea turtles, she then talks about the problem with nurdles. Turtles and nurdles: it all sounds like something from a Dr. Seuss book. This interest in nurdles—which, as it turns out, is the term for preproduction microplastic resin pellets— first comes as a surprise to me. But when one takes into account the simplicity of turtle-tracking tasks (sea turtle nests being few and far between, after all), and the volunteers' motivation to make a difference, it seems inevitable that they would find another way to realize their full volunteering potential.

---

[79] Wrightsville Beach is a link in the chain of barrier islands parallel to the coast.

Sea turtles spend only a few key hours of their entire life away from the sea, going only as far as the high tide mark to lay their eggs, but during their brief time on land they face terrestrial threats, such as raccoons, which eat turtle eggs. This, coupled with pollution in the ocean, has knocked down sea turtle populations. The loggerhead sea turtles live in the Atlantic, Pacific, and Indian Oceans, nesting on many coasts. To recover this global species involves making specific plans for each population.

Populations are delineated based on where females nest; there is no other way to pin down their national allegiance. Peninsular Florida and Masirah Island, Oman, are the two largest nesting areas, each with more than ten thousand females every year. Loggerheads nest in lower numbers in Georgia, South Carolina, and North Carolina; this region is covered by the Northern Recovery Unit. After hatching, the hockey puck-sized turtles from the Carolinas and Georgia make their way to the Mediterranean Sea and only return when ready to breed. In the deep ocean, they live in beds of free-floating seaweed. After a decade or so, when they reach the size of a Frisbee, they move into shallower shore areas, called neritic habitat, where the water depths do not exceed seven hundred feet. Other loggerheads nest in Japan and spend their youth off the Mexican coast; still others nest in eastern Australia and grow up along the coast of Peru. Young turtles ride the prevailing ocean currents much like human youth ride megamall escalators between stores.

Miller grew up in Ohio and has no immediate plans to float home as a turtle would; she loves the beach as an ecosystem. More common attitudes toward the beach are to treat it as a romantic getaway, a playground, even a giant ashtray. Beaches in the early morning resemble a fraternity house's lawn on the morning after. Only when I am able to adopt Miller's view and see the beach as habitat, am I able to fully appreciate that abandoning a plastic bucket in the sand is not the bratty but otherwise benign act of a child refusing to put away her toys, nor merely littering, but an insidious form of pollution. Maybe the heat has made me cynical, but I decide here and now that beachgoers are slobs.

A bizarre news story telling about a long-deceased person found in an apartment buried under piles of debris is no longer unfathomable; this is surely a potential fate of beachgoers.

Miller is good-humored in the heat and filth. Through sweat-and-sand-encrusted eyelashes she scans the sand at her feet, picks up three partially buried straws and a bottle cap and chimes, "I can skip the gym today, I did my squats."

All species of sea turtles, including the loggerhead, are on the International Union for the Conservation of Nature's Red List of Threatened Species, along with charismatic species like Siberian tigers and Black rhinos. Sea turtles are rarely seen, but they garner public support. They represent characteristics that we value: their silence and gentle determination, wrapped in a beautiful armored shell, are forcefully inspiring. They are, as turtle geneticist Brian Shamblin puts it, cosmopolitan, the foremost world travelers.

Most well-traveled species are thoroughly mixed, genetically, so that all individuals are cousins and it becomes impossible to parse out populations among one big extended family. Turtles are an exception, because females have fidelity to their birth areas. Some flow of genes between populations still happens through choosing mates, when female turtles mix with fellas offshore who may or may not be local. Thus, Shamblin can only define the geographic boundaries of rookeries by the females present. He uses DNA from the mitochondria of cells rather than from the nucleus of cells; mitochondria are handed down from mother to daughter to granddaughter—like pie recipes.

In 2010 the National Marine Fisheries Service crunched numbers from genetic, demographic, geographic, and oceanic factors and identified nine major sea turtle rookeries. Evidence suggested that the North Pacific and South Pacific populations were distinct and that both should be listed as endangered, while the other seven populations (the Northwest Atlantic, Southwest Atlantic, Northeast Atlantic, Mediterranean, Southwest Indian, Northwest Indian, and Southeast Indian) continue to be designated as threatened. In 2014 Shamblin completed a more in-depth study of the DNA from six of the nine regional rookeries. His findings support the distinction between the populations, as well as identifying eighteen other subpopulations that are unique enough to warrant their own management strategies. The subpopulations within the populations were again a surprise to find in such a cosmopolitan species. This goes to show the power of sisterhood.

After delineating populations, the second order of business under the Endangered Species Act is for the federal agencies to identify critical areas of habitat for the nine populations, such as safe areas for foraging, resting, and nesting. Because loggerhead sea turtles use terrestrial and marine habitats, critical habitat needs to be identified by the US Fish and Wildlife Service and the National Marine Fisheries Service. Both did so for sea turtles in 2013. The Fish and Wildlife Service proposed critical habitat for Northwest Atlantic loggerhead populations that included 739 miles of coastline from North Carolina to Mississippi. That's 84 percent of all nesting habitat for loggerheads in United States. The National Marine Fisheries Service proposed thirty-six marine areas for the same populations; these included neritic areas, breeding areas, foraging areas, wintering areas, migration pathways, and open deep-water habitats with floating beds of seaweed.

Walking on the debris-festooned beach at dawn makes me wonder what pleasantries must float among the nursery school turtles living in the oceanic beds of seaweed. On the sand we collect diapers, diaper wipes, abandoned plastic sand toys, cigarette butts galore, plastic straws, plastic grocery bags, plastic forks, balloons, soda cans, soda bottles, T-shirts, underwear, condoms, tampon applicators, plastic lotion bottles, fast food bags, tiny plastic condiment sacks, plastic water bottles, and many unidentifiable objects of plastic, metal, polystyrene, and wax paper. When the tide comes in, these sins are washed away with the sand castles.

About 90 percent of the trash that Miller and the other volunteers collect is made of plastic. All plastic products are created from the tiny beads called nurdles. Before I can understand how plastics are produced, I puzzle over why plastics are strewn across the beach. There are clean blue garbage bins at every access point, about forty feet apart. Every person who left the beach on foot walked past a garbage bin. Are beachgoers litterbugs? Worse is that when the team of volunteers first went public with the amount of garbage they collected, people scoffed in disbelief, saying the beaches look clean enough to them. Ah, so say people who have accepted living with sand in their every nook and cranny. Miller, a generous person, explains the incongruity in perceptions. When the beach is crowded, it is easy to assume that everything scattered about the sand belongs to someone in the sweltering throng. The trash is camouflaged as property. Only in the early morning, when the beach is deserted, can one see the extent of jettisoned materials that go unclaimed. Miller doesn't let me put our spoils into the blue bins. The protocol is to take it all home, rinse it, sort it, and photograph it. "That's the only way we could convince people that we weren't liars," she explained.

Miller and I collect the equivalent of five bags of groceries, which she will take home to rinse, sort, and photograph. We have not found any turtle tracks during my visit. If it weren't for Miller's efforts, we wouldn't know that Wrightsville Beach doesn't get many turtle nests; indeed, sea turtle conservation is only possible because of the observations by volunteers, like Miller, who are up at dawn to patrol their assigned three-quarter-of-a-mile stretch of beach looking for nests.

Volunteers who collect DNA, which comes from eggshells, are pivotal too. These volunteers have special permits and additional responsibilities. Like turtles, the volunteers have been delineated into groupings, in this case based on the stretch of beach they monitor. The nesting beaches in North Carolina are divided into twenty-two geographic units, sixteen of which are monitored by volunteer groups. Each volunteer group has an appointed coordinator who oversees the other volunteers in the group; the coordinator collates all the monitoring observations and provides the data to state biologists, and also receives a permit to do a wide range of special tasks. For instance, it is the coordinators who place signposts at nests: four wooden stakes connected by flagging and a sign from the North Carolina Wildlife Resources Commission that explains why the area is staked and warns of the law and the penalties for disturbing the nest, eggs, or hatchlings. Coordinators are allowed to hang netting around nests in order to block outdoor artificial lighting from illuminating nests. And coordinators are permitted to collect eyeballs from dead, stranded turtles.[80]

The head volunteer for Wrightsville Beach is Nancy Fahey, and she employs a divide-and-conquer strategy with her volunteer assignments. Sea turtle monitoring is subdivided into three-quarter-mile sections, just one of which I walked with Miller. As the coordinator, Fahey assigns seven volunteers (one for each day of the week) to each section, rotating them so that every section is monitored by 8:00 a.m. every day. When a volunteer finds evidence of a nest, he or she contacts Fahey, who meets the volunteer at the site and determines whether it is real nest or a false crawl (sometimes females come ashore but then do not lay eggs). It is Fahey who has the necessary permits and training to remove and preserve one fresh egg from every nest in her jurisdiction. She will discard the yolk, place the leathery shell in vials full of preserving solution, label the vial with date, location, and nest identification number, and send it to the lab.

The lab is part of a multistate project among agencies for North Carolina, South Carolina, Georgia, and the University of Georgia to figure out the number of loggerhead sea turtles that nest in the region. The lab uses DNA fingerprinting to identify individual nesting females, how many nests they lay each year, and how long they go between nesting years. The data will make a strong census and create a better picture of population dynamics. Coordinators across the three states have collected eggs from 38,459 nests. The lab has been able to retrieve DNA from 85 percent of those, which has led to the identification of over seven thousand unique females. Females had, on average, 4.5 nests per season, and took about nine days between nests, which were usually within twenty-five kilometers of each other.[81] Shamblin found a beach with three generations of loggerheads nesting simultaneously: a grandmother, her daughter, and her granddaughter.

---

[80] For science, of course. Studies of turtle eyes are used to understand turtle growth.
[81] One female laid nests as far as 650 kilometers apart, with the first nest in Cape Lookout National Seashore in North Carolina and about a month later a nest in Little Cumberland, Georgia. She likely also laid a nest or two in between.

Conservation doesn't succeed if we only study the delineation of populations, estimate their size, and designate critical habitat. Even if the critical habitat is protected from development on land and from fishing at sea, there are still more hazards. Managing turtles really means managing humans. Case in point: litterbugs. Trash isn't just unsightly; it is bad for the health of turtles and humans. The idea that human health is tied to the well-being of wildlife and the environment is called One Health.[82] Before we pick through the issue of trash, let's shine the spotlight on two other enemies of the sea turtle: nighttime light and sunbathers.

Despite their ability to navigate the global oceans, the fate of sea turtles can have a simple twist by a poorly timed evening barbeque illuminated by tiki torches in the backyard. When turtles are on land, they navigate by instinctively moving toward the brightest light close to the horizon. They aim both eyes to receive equal amounts of light, a behavior called phototropotaxis. In the pristine intertidal world that has existed since the Triassic period, over two hundred million years ago, the brightest light at night was the moon or starlight reflected on the ocean—that is, until Thomas Edison invented the light bulb. Water reflects more celestial light than land, so sea turtles would always find their way back to the sea. Now with a flick of an outdoor light switch, newly hatched turtles can be misdirected and kept from finding their way to the ocean.

When the will is present to protect turtles from light pollution, the way is pretty obvious. Coastal communities with volunteer sea turtle projects have usually passed local ordinances to limit outdoor lighting at night. When it comes to the desire for natural darkness, this issue unites those with interests in wildlife conservation and astronomy. Advocating for dark skies is a battle against the human primordial fear of the dark. Our childhood cry for nightlights grows into demands for street lighting that functions to reduce feelings of fear—particularly fear for personal safety among women. Studies bear out the idea that crime is reduced by street lighting, but not as one would expect. Light-

[82] The notion of One Health was conceived after the 2002 Ebola outbreak.

ing is a placebo that everyone, even perpetrators, swallow. Lights don't deter crime by increasing surveillance ability; when lighting is present, nighttime *and daytime* crime is reduced. Researchers speculate that lighting strengthens social pride, confidence, community cohesion, and social control of neighborhoods, and these social factors operate to reduce crime.

Wrightsville Beach is a fairly tight-knit community with about 2,500 year-round residents. The town passed an outdoor lighting ordinance in 2012 and the law is as much on behalf of the sea turtles as people preferring the ambiance of a small town instead of the brightness of resort cities like Las Vegas.

On Wrightsville Beach, as with everywhere else, controversies begin when money is at stake. Our second villain enters the scene amid a continual tension between businesses in the tourism industry and full-time island residents. The former wants the tax dollars of the latter to be used to protect island infrastructure. All infrastructure—bridges, roads, buildings—requires stable beaches. Of course, sunbathers need beaches too, which you might think are in plentiful supply. They aren't. Beaches are deceptive; they appear calmly predictable with the rhythmic motions of waves, but in reality, they are like restless children, constantly in motion and causing trouble. People want beaches to behave, like the mature mainland: the tectonic plates of the mainland move anywhere from zero to one hundred nanometers (one billionth of a meter) per year, but the Outer Banks erode at an average rate of about two to three feet (one-half to one meter) per year. Sands wash away from some areas and pile up in others. The losses are greater than the gains. One study has found that these barrier islands have experienced 70 percent erosion and only 30 percent accretion.

Of course, natural disasters bring quicker change. When tectonic plates nudge one another, like teens elbowing for more room, they produce volcanoes and earthquakes. The rage of one hurricane can help the sea claim a beach and everything on it. Even putting natural disasters aside, constructing a building in the middle of a barrier island is akin to placing it on a moving conveyor belt. The building will eventually be at the edge of the island and, one day, off the edge. Retirement homes, vacation retreats, and other amenities for tourists are all at risk from the high erosion rates of beach ecosystems. Visiting the beach might be fun for everyone (except me), but living on the beach is a constant struggle against the forces of nature. Tourist is a season after all, not a permanent way of living.

Sea turtles can handle the dynamic and moody face of the coastline, and beaches have acted as the birthing centers for sea turtles since the dawn of their time. Sunbathers and their ilk, on the other hand, are full of ways to counter nature from taking its course. It goes without saying that attempts to legislate nature into compliance don't work.[83] Other countermeasures work in the short term but have unanticipated consequences. For example, who among us would have anticipated that adding stability to beaches would alter the sex ratio of sea turtle offspring?

The tourism industry puts municipalities under pressure to protect resorts from sliding off the conveyor belt. The old solutions of sea walls and breakwaters displaced problems from one municipality to another. These are outlawed now; you just can't fool Mother Nature that easily. The alternative interventions are beach nourishment projects. Nourishment costs millions of dollars, and one hurricane or tropical storm can undo the work overnight. Wrightsville Beach is "nourished" every four years.

Beach nourishment involves dredging offshore sediment and pumping it onto eroding beaches. "Native sand is like sugar," Fahey explains, describing the texture of pristine beaches. Nourished beaches have shell hash and sand with coarser grain. More important, dredged and pumped sand is a different temperature from the original beach sand.

Like other reptiles, sea turtles have temperature-dependent sex determination. The temperature of the sand, rather than a chromosome, determines the sex of each hatchling. Warmer temperatures produce fe-

---

[83] In 2012, the North Carolina General Assembly passed a law forbidding communities from passing local laws and ordinances based on the most recent climate change predictions about sea level rise. Ignoring sea level rise won't prevent nature from following its own laws.

males; cooler temperatures produce males. A diversity of incubation environments across the loggerhead range is important to attain a healthy balance in the proportion of males to females. Historically, North Carolina beaches produced large quantities of male hatchlings for the greater loggerhead sea turtle population. Beach nourishment adds warmer sands; consequently, eggs now produce a greater number of females.

In addition to periodic beach nourishment, nature has been sporadically diverted from its course in bigger ways. Back in the day, the area now called Wrightsville Beach was called New Hanover Banks in the south and Shell Island in the north; the two were bisected by Moor's Inlet. At some point, Moor's Inlet was bulldozed over, and the northern border became Mason Inlet (the southern border is Masonboro Inlet). There were not many visitors to the island until 1887, when a turnpike was completed that connected Wilmington to Wrightsville Sound. The turnpike was topped with oyster shells and dubbed Shell Road. The area's heyday for tourists followed. Still, the sands changed faster than the names. More than a century later, Mason Inlet started aggressively migrating south, bringing Shell Island Resort to the very edge of the conveyor belt. In early 2002, engineers created a new inlet (unsurprisingly called New Mason Inlet) about three thousand feet to the north, and then filled in Mason Inlet completely to protect the resort. While some people are metaphorically moving mountains to protect sea turtles, others are literally moving beaches in ways that threaten them.

One of the biggest threats to sea turtles is not easily addressed by monitoring the turtles themselves; it is addressed by monitoring trash—in particular, the plastics made from nurdles. When I joined Miller in gathering the garbage that she documented in photographs, this was independent of her volunteering for the state's sea turtle program. When I first spoke to the turtle volunteer coordinator, Fahey, on the phone to arrange a visit, she warned me, "We are a small-time operation"; she meant they didn't find many turtles on their five-mile stretch of beach. I justified the visit because, as far as science goes, recording the absence of turtles is as important as recording their presence. It was a surprise to me to learn that Fahey's "small-time operation" had spawned its own citizen science project that studies beach debris. Wrightsville Beach—Keep it Clean is the brainchild of a highly spirited volunteer, Ginger Taylor, who started the project to satisfy her own curiosity about the amount of litter on the beach. She continued the project to bring awareness to the issue and advocate for solutions. As the data collection grew, so did Taylor's involvement in attending town meetings and working with community members and businesses for change.

The formation and impact of Keep it Clean are the most mysterious, unpredictable, and amazing types of outcomes for citizen science. It is great to design projects for environmental justice, to manage species and habitats, and to solve community problems. It is *phenomenal* when a completely contributory project—as the turtle monitoring designed by the state government—spawns a community-based action project. The problem with plastics would have remained hidden from community view if a fairly traditional project (monitoring turtles) had not put extra pairs of alert eyes on the beach, brought together activist types of people, and encouraged people to really be aware of their surroundings. A heightened awareness, combined with efficacy for science and monitoring, are the ingredients for change.

Taylor, a licensed clinical social worker by day, hatched the idea for Keep it Clean at night, while sitting with a small group waiting for sea turtles to hatch. Beach patrol to find nests begins in May and ends in mid-September. Another task for volunteers is sit duty, which begins in late June and ends by Thanksgiving. Sit duty involves arriving at 6:00 p.m. and waiting in the dark for eggs to hatch, sometimes until 11:00 p.m. Most things that get buried underground —such as dog bones, pirate booty, and corpses—rarely resurface, at least not of their own volition. Sea turtle eggs are more like flower bulbs: they bloom with young turtles, all at once, when the time is right.

About two months prior to this egg emergence, while people slumbered, a miniature armored tank of a turtle planted her eggs in a pit in the sand after she slowly pulled herself on to the beach. It took her about an hour to dig a pit above the high-tide mark and drop over a hundred white, leathery Ping-Pong-ball-size eggs into her hole in the sand. (A turtle produces no body heat herself, so the eggs are better off in the sunbaked sand.) After covering the eggs, she lumbered back to the sea, leaving the eggs forever. Ectothermic reptiles have a cold-blooded maternal strategy.

After about two months the turtles hatch and wait in the darkness of the pit until nightfall. When the sand cools after sunset, the baby turtles take that as a signal to bloom. For sea turtles, darkness is safer because predatory seabirds are asleep. Turtle hatching is called a boil, which in my interpretation is because the young bubble up to the surface of the sand like the brew in a witch's cauldron. "A watched nest never boils," said no sea turtle volunteer ever. Volunteers who sit for a boil look like they are sitting at a campfire, mesmerized by flames, but instead are staring at a patch of sand demarcated as a nest by the wooden stakes and cautionary tape. When baby turtles start crawling out from the pit in the sand, it looks like the swarming return of the undead: I would expect the monitors to feel a touch of horror as though hundreds of tiny, spiky prehistoric reptilian vampires were clawing their way out of a mass grave. Instead, they pack the faintest of whispers with the emotional intensity of Beatlemania screams. Single-mindedly, the little turtles crawl like infantrymen across the beach until the surf sweeps them away.

Three days after the hatch, the volunteers return to the nest to excavate it. They dig out all the eggshells and any remaining eggs, then count and record how many hatched and how many didn't. Sometimes a hatchling will be trapped in the nest pit and the volunteers will set it free. At Wrightsville Beach, they invite as many people as possible to witness the excavations and share in the joyful miracle of sea turtle afterbirth.

The first time Taylor saw people at night, huddled up at the beach without a campfire, she thought they were having a séance. That was in 1998 and she stopped, learned about turtles, and asked a million questions, but she could not commit to volunteering because of her work schedule. Nine years later, while working a contract job in Hawaii, she became enchanted by green sea turtles that would come onto the beach just north of Haleiwa to bask in the sun. When she returned to Wilmington, she was ready to join the efforts at Wrightsville Beach, patrolling in the early mornings and sitting in the evenings. She saw her first boil in 2007 and described it as magical and spiritual—a natural high: "It puts things into perspective." While the nights were amazing, the morning patrols were not. At first she and her husband imagined leisurely romantic strolls on the beach together. But she couldn't walk by the trash without picking it up; Taylor isn't the bystander type.

It was one evening while sitting with others and waiting for a nest to boil at Wrightsville Beach that Taylor raised the issue of the trash. She quickly learned that many other volunteers were already taking the initiative to pick it up. She wondered how much they had collected, and her curiosity propelled her and the other volunteers to start a new citizen science project. From the turtle program, they were already familiar with protocols to standardize and collate data collection and the subsequent sharing of the information: "We got this." Taylor was confident they could do something similar with trash.

Her plan evolved over time. In the early days of quantifying trash, they called themselves the Trashy-Talking Turtlers. While Fahey was compiling nesting data from volunteers to report to the state, Taylor was compiling data on the number of trash bags filled and reporting it in a weekly newsletter that she sent to participants and town officials. Staff at the local paper, the *Lumina News*, read everything sent to town officials, and so they picked up on the story from time to time. As public discussions ensued about how to address trash on the beach, people started to question how their tax dollars were spent, which in turn led to some skepticism about the trash, and suggestions of junk science that amounted to calls of "show me the garbage." Wrightsville is a beautiful, beloved beach and many beach revelers are blind to the trash. Several volunteers noted that the garbage is hard

to notice when the beach is crowded and that it takes a special inten-
tion, and trained beach-combing eyes, to spot garbage in the sand.
Taylor brought bins upon bins of saved, rinsed-off garbage she gathered during her patrols and presented it at a board meeting. She didn't
enjoy discrediting claims that Wrightsville Beach was pristinely clean
—that's why people love it—but she could not ignore her evidence.
She had to open their eyes to the sad truth, believing that people will
protect what they love and knowing that the people of Wrightsville Beach
truly love the ocean.

"When the trash issue grabbed public interest, we decided to separate from the turtle monitoring," Taylor explains when we meet in
an air-conditioned coffee shop in Wilmington. "Plenty of people without particular interest in turtles had big concerns about the health of
the oceans and we wanted the name of our group to reflect the
broader interest. Plus, turtles had nothing to do with the garbage
problem, and we didn't want to risk getting them dragged into the
potential controversies." Same beach, same morning patrols. Some of
the same turtle people, plus more. In 2010, the group gathered name
suggestions and took a vote and agreed on Wrightsville Beach—Keep
it Clean. Taylor continues to assemble the data on the amount collected from each zone and how frequently each zone is cleaned. The
town board has used the data to help address litter problems, including changing some ordinances.

The public skepticism taught the volunteers to not throw the evidence out with the trash, even when the evidence *was* the trash, until
first documenting it in photographs. One could view the public response as a form of peer review. As scientists learned long ago, purely
motivated skepticism expressed through peer review is an effective way
to improve research projects.

No one, least of all Taylor, anticipated what happened next. She had
gained a new perspective from watching turtles hatch and, similarly,
she gained a new perspective by looking at the photographs of the
sorted refuse. She saw a prevalence of plastics, and this observation led
to more questions. That's how science operates: answers lead to more

questions and the cycle perpetuates itself, just as a child maintains a circular conversation with an adult by responding, "But why?"

Taylor turned to Google and searched for "turtle" and "plastic." She learned of the Great Pacific Garbage Patch, which is a soup of microplastics spanning twice the size of the state of Texas; the two major areas of concentration are the Western Garbage Patch, off the coast of Japan, and the Eastern Garbage Patch, situated between California and Hawaii. She learned that because plastic items are made of crude oil, they never decompose, but they do slowly break into smaller (sometimes tiny) pieces due to exposure to sunlight. She learned of ocean garbage patches in other parts of the world and the impact of plastics on marine life. Like accidentally viewing a grisly image that you wish you could unsee, Taylor saw an ugly global problem and could not ignore it. The data on trash from Keep it Clean didn't lead her to discover something previously unknown to humanity, but for her and the other volunteers, finding out that their beach was part of a known global problem made this a very personal discovery. And personal discoveries can be a force for change.

The mass-production of plastics began in the 1940 and 1950s, and today over 260 million tons of petroleum-based goods are produced every year, among them pipes for household plumbing, patio chairs, intravenous drip bags and syringes, computer keyboards and mice, and polyester shirts. About half of what's produced will be disposed in landfills or recycling centers *within a year*. The rest is unaccounted for. Much of it may remain in use, but some crazy, unknown percentage has been added to the piles of litter that cover our beaches and float in our oceans. The top three generators of plastic waste are the world's population centers: China, the European Union, and the United States.

Plastics are moldable organic polymers, derived from crude oil. Crude oil is refined into products like gasoline, diesel fuel, kerosene, and heating oil, and, a little farther down the production line, into crayons, bubble gum, tires, and spandex pants. Crude oil is made up of hundreds of different types of hydrocarbons, which are hydrogen and carbon atoms bonded into various sorts of arrangements. The industrial refinement process segregates hydrocarbons based on their boiling points.

Different hydrocarbons vaporize at different temperatures. Refinement starts with a slurry of hydrocarbons. As each vaporizes, the resulting gas is funneled into a chamber to cool and condense. Then the segregated hydrocarbons are combined to make compounds like ethylene, styrene, and propylene, among hundreds of others. Because there are so many types of these compounds, they are generically called monomers. To design different plastics, one mixes and matches various monomers into polymers. Nature does the same; the difference is that natural polymers, like cellulose and vegetable oil, will biodegrade because they are derived from living things. Polymers derived from petroleum will only photodegrade. Polymers are further designed through the addition of chemicals to make them soft and pliable, dyes to make them colorful, and flame-retardant chemicals to suppress combustion. Over half of all plastics are designed with chemicals that cause cancer, and some with chemicals that mimic hormones (we'll return to this topic in chapter 10). The final juiced-up polymers are shaped into the tiny resin pellets called nurdles, the building blocks of plastic goods.

The making of toothbrushes, iPhone cases, milk jugs, and teething rings begins with nurdles. Heated and molded, nurdles are quite like colorful Fuse Beads, a popular craft that kids arrange and their parents iron into place. Given how they are made, is it dangerous to use plastics?

In 1998, Hideshige (Shige) Takada, an organic chemist with expertise in measuring trace levels of chemicals in the environment, was approached by one of his colleagues and asked to examine some nurdles.[84] The expectation was that nurdles would have only trace levels of hazardous chemicals, since only trace amounts of known carcinogens are

---

[84] In his early career, Takada studied polycyclic aromatic hydrocarbons (PAHs) in urban street dust, and then tar-balls on the coast. He followed pollution in runoff, all of which eventually ends up in the ocean. Coincidentally, my husband, an astronomer, studies PAHs emitted by very old stars.

added to polymers. Furthermore, pure plastic is expected to have low toxicity because it doesn't interact with its environment. It is what chemists call inert. Inert chemicals are uncompromising, showing no give-and-take with their surroundings. The expectation was wishful thinking.

Takada found that concentrations of organic pollutants are incredibly high in nurdles. The pellets are, by some standards, biohazards, chock full of persistent organic pollutants (POPs).[85] He figured out that nurdles in seawater are not inert, but bond with a lot of chemicals diluted in the sea, such as PCBs (toxic by-product of industrial production), DDT (an insecticide), and PBDEs (flame retardants). Even the plastics designed without added toxins quickly become toxic. For instance, polyethylene, used to make plastic bags, is probably not toxic in your home,[86] but it becomes toxic by absorbing pollutants, which happens when the bags float around the ocean. That might be a good thing if the nurdles and other plastics were then sieved out of the ocean, but they're not; instead, they enter the food chain.

Once they invade the lower levels of the food chain, they bioaccumulate their way to the top. In one study, seabirds that had eaten plastics had polychlorinated biphenyls in their tissues at 300 percent greater concentrations than those that had managed to avoid plastics. POPs are found on plastic waste in concentrations one hundred times higher than those in sediments and one million times higher than those occurring in seawater. Many POPs are regulated by the US Environmental Protection Agency (EPA) because they disrupt cell division, affect immune functions, and cause diseases. Almost 80 percent of EPA priority pollutants are associated with plastic debris (by virtue of being designed with hazardous ingredients or by gaining the hazards from the environment). Taylor sums up the situation simply: "It makes you cry." Just as people do not view the beach as habitat, humanity does

---

[85] When talking about chemicals in the environment, the word *organic* has a very different meaning than when talking about food. Organic, to a chemist, mean "molecules that contain carbon."

[86] That's good news for my cat. She likes to lick these bags.

not view the ocean as a living ecosystem. It functions as a cesspool, and every high tide is the swooshing flush that fills a couple of Texas-size toilets.

Takada had found that nurdles concentrate POPs so highly that just five of them have the equivalent POPs of twenty-six gallons of seawater. This calculation led him to realize that nurdles provide a simple way to monitor global contamination of POPs. Now it made sense for him to start a citizen science project. He could never ask people to send twenty-six gallons of seawater to him, but he could ask for microplastic pellets. In 2005, he launched International Pellet Watch, which asks people to look for nurdles on the beach and mail them to him for chemical analyses. Volunteer contributions have led to the identification of the more and less polluted areas of the world. For instance, nurdles found on beaches near big cities such as London, New York, and Tokyo have skyrocketing concentrations of PCBs.

Nurdles spilled at production sites are one source of microplastic pollution. Another is that big pieces of plastic don't stay together forever. Plastics break into smaller and smaller pieces over time. Eventually they reach a size where fish, invertebrates like zooplankton, and microorganisms ingest the plastic.[87] Wood, leather, glass, and clay can easily return to be part of the earth. Plastics haunt us forever.

Taylor has two sobering words for me: "endocrine disruptors."[88] The chemicals that leach from plastics interfere with hormonal systems, which are major channels for communication among various parts of the body. Taylor tells me one example of endocrine disruption that she learned from Charles Moore, author of *Plastic Ocean*, a book that describes his experiences with garbage patches in the ocean: "When mice ate particular plastics, they lost their maternal instincts." All species of sea turtles, and at least 21 percent of all seabird species,

[87] Many sea animals are unable to distinguish garbage from food. Ingesting garbage can result in a blockage of their digestive systems, and many animals thus starve to death. Others survive just long enough to contaminate the food chain.
[88] In the iconic 1967 film *The Graduate*, a family friend advising Benjamin (Dustin Hoffman) on possible career pursuits proclaims, "I just want to say one word to you. Just one word . . . Plastics."

are harmed by plastics. People are harmed too. Plastics are either made with chemical toxins, or in environments where they absorb toxic pollutants.

Since plastics are toxic, why are they treated as solid waste? In a 2013 article in *Nature*, marine scientists and chemists (including Takada) urge countries to regulate plastic waste as hazardous materials. They argue that doing so would quickly help agencies restore affected areas and be impetus for research on new and safer polymers. Four plastics make up 30 percent of all that is produced: PVC (in pipes that carry our water), polystyrene (in food packaging), polyurethane (in furniture), and polycarbonate (in electronics). These are difficult to recycle and sit as toxic time bombs. If just these four types of plastics were categorized as hazardous, in the United States the EPA could follow the Comprehensive Environmental Response, Compensation, and Liability Act of 1980 to initiate a massive cleanup effort.

Ever since that fateful day of Googling, Taylor and many of the other volunteers reduced their use of single-use plastics, choosing alternatives whenever possible. Eliminating some, like straws, was easy. Vanquishing others, like sporks, was a pleasure. Yet plastics are so pervasive that it is a challenge to avoid them, like zipper-lock food storage bags and adhesive bandages. In the coffee shop, I sit with a smoothie in a single-use plastic cup with a single-use straw; Taylor has given her reusable canteen to the barista to fill directly. I think part of the difference in our behavior speaks to the power of experiential learning, which often accompanies participation in citizen science. A key part of learning is the process of making meaning, and we make it best if we can base it on firsthand experience.[89] I learned about marine pollution in the academic literature and news articles; Taylor learned from direct contact and her own inquiry.

Another way to explain the heightened responsibility among volunteers like Taylor, Miller, and Fahey is that they are data driven in

[89] Taylor also had the experience of holding a sea turtle at a rehabilitation hospital while its tank was being cleaned, and it defecated a red latex balloon. In databases of turtle necropsies, many note balloons found inside them.

their decision making. They are not that different from people with diabetes who learn to track their insulin levels in order to successfully manage the disease. Today people are using new technology to gather data on vital signs, cortisol levels, how many steps they've taken, what they've eaten, the medicines they take, and their personal mood, weight, posture, and sleep patterns. This is part of a personalized health movement called the Quantified Self. Gone are the days of a simple pedometer; these are the days of Fitbits and Bioharnesses. The data usually make their way to marketing companies, but setting that aside, when people have data in their hands, they can use it to guide decisions about diet, exercise, medicines, friendships, and more. In essence, Taylor, Miller, and Fahey are managing their lifestyles and environment and using data on trash to inform their personal management decisions. Plus, quantifying garbage informs the decisions of police (they can increase patrols of highly trashed areas), town boards, and members of the public.

The sea turtle volunteers appear to be immune to a social and psychological phenomenon called the bystander effect, where a person is *less* likely to assume moral responsibility if others are present to (potentially) assume it instead. Yet even while assuming responsibility by picking up other peoples' trash, the volunteers talk about the need for industry to do its part by assuming responsibility for producing social good.

Miller equates industry's avoidance of responsibility with the ubiquitous habit of passing the buck that we teach our children. She had moved to Wilmington from Ohio to become a public school teacher. In Ohio, competition for teaching jobs was fierce; while aggressively pursuing one job with multiple follow-up calls, she was told to back off: "Honey, you are one of nine hundred applicants." Meanwhile, North Carolina had a teacher shortage. She was instantly hired over the phone with "School starts in two weeks; can you be here?" Three years later she quit. With the passing of the No Child Left Behind Act, she notes, "A teacher really can't give a failing grade. The teacher has been made responsible for passing kids, rather than the kid being responsible for earning a passing mark. I couldn't stay part of that system."

Taylor urges others to accept more responsibility, one company at a time. Several months earlier before our meeting she had gone into a local market for her socially responsible shopping. The local honey supplier was using plastic bottles. Taylor called the supplier and left a heartfelt and educational earful via voicemail. Although she never received a call back, the supplier had gotten the message: several months later Taylor noticed that the local honey was now in glass bottles.

Keep it Clean joined with the Plastic Ocean Project and Cape Fear Surfrider on a campaign called Ocean Friendly Establishments that encourages, certifies, and supports local restaurants that only provide straws upon request. Straws are consistently among the ten top litter items in annual International Coastal Cleanup reports. One of the biggest resorts, the Blockade Runner, was the first to sign up for the program with its East Oceanfront Dining restaurant.

I first thought of Miller, Fahey, and Taylor as midwives for sea turtles, managing the hatches. They participated in a citizen science project run by a state government agency, and their essential roles gave them leverage to be decision makers in the management of an endangered species. This empowered them to set an agenda for their own citizen science project, which transformed their lives. I now see them as stewards of turtles, a role that extends their responsibility from the local beaches to faraway oceans.

Science is our most reliable system for obtaining new knowledge quickly. For any given research question, if we do science properly, we should come to the same answers. Thus the most influential part of science is selecting the most important and relevant research question. No one can know in advance what the answer will be (science is, by definition, discovering something not yet known). But who selects the question is important to many types of citizen scientists. Who decides which knowledge to pursue? The volunteers on Wrightsville Beach initially came together over a common interest in turtles, and from their shared experiences they developed their own citizen science project. In chapter 9 we'll explore how citizen scientists take further control to improve lives and environments.

CHAPTER 9

Geography
White Picket Fencelines

First they ignore you, then they laugh at you, then they fight you, and then you win.
—MAHATMA GHANDI

"LOCATION, LOCATION, LOCATION" IS THE REAL ESTATE AGENT'S MANTRA. The central importance of location is also the core tenet of the field of geography, which is the study of places and the relationships between people and their environments. Geography is spatial in the way history is temporal: it is about understanding the physical relationship among places instead of the relationship of events along a timeline. We can analyze the past to understand how historical events influence the present. Similarly, we can assess one place and understand how it influences a neighboring place. In this chapter we'll visit three places influenced by their immediate neighbors: Tonawanda, in upstate New York, with a neighboring coke factory; parishes in New Orleans, Louisiana, with neighboring petrochemical oil refineries; and Tillery in eastern North Carolina, with neighboring industrial hog farms. To activists supporting environmental justice, these communities are known as fenceline communities. Instead of the proverbial neighbor behind a white picket fence, these communities have neighbors who claim corporate personhood and cast a shadow of industrial pollution over the fence. Only when residents measure pollution are they able to succeed in legal, regulatory, and political actions that protect their communities.

Most of the citizen science projects we've examined thus far were initiated by scientists and engaged hundreds, thousands, and sometimes tens of thousands of laypeople. These citizen science projects built scientific knowledge that would be otherwise unattainable, and scientists published the discoveries in scientific journals. When citizen science is initiated by residents in communities facing local problems . . . well, let's take a closer look at what happens.

In the blue-collar town of Tonawanda, New York, many residents were fighting for their lives against rare cancers, and almost everyone was dealing daily with skin rashes, burning mouths, and pungent smells. Residents described the air as having a foul, acrid, tar-like stench, strong enough to wake them in the night to shut the windows. While most of us check outside for hints of rain, residents in fenceline communities check the wind direction to decide if it is safe to go outside. To figure out what was causing the air to be bad, Jackie James-Creedon, Adele and Bob Henderson, and Tim Logdson learned how to build an air-sampling device made almost entirely from materials available at a building supply store.

After dark on August 16, 2004, James-Creedon, the Hendersons, and Logdson gathered about a half a mile downwind of Tonawanda Coke Corporation to collect a sample of the air. They recorded details of the environment: there was a light breeze that measured at six miles per hour, and the temperature was a comfortable sixty-four degrees.

The low-cost bucket design of their air-sampling device was originally crafted by an environmental engineer hired by Edward Masry, the crotchety litigator famed for wining a class-action suit against Pacific Gas and Electric that awarded over $300 million to victims of chromium poisoning in Hinkley, California; the story of the suit was made into the Oscar-winning Hollywood blockbuster *Erin Brockovich*. In the air-tight bucket was a special Tedlar bag purchased from a chemical analytic company. James-Creedon and her gang pumped the air surrounding the bag out of the sealed bucket to create a vacuum. At 10:50 p.m. they opened a valve and air was sucked into the bag as it expanded into the pressureless vacancy within the bucket. The night air carried a heavy mix of tar and ammonia odors, so thick they could taste it. At 10:53 p.m., they closed the valve. They completed chain-of-custody paperwork, which served as the official document turning the air sample into physical evidence. When they opened the bucket, the Tedlar bag was puffy like a clear pillow; they prepared the bag for shipment to Columbia Analytical, a lab in California, and then they went home, each with nausea and headaches.

Columbia Analytical is one of the few companies with special apparatus that allows it to attach the Tedlar bag to a gas chromatography mass spectrometer. The sample from Tonawanda whistled with benzene at a concentration of fifty-four micrograms per cubic meter ($\mu g/m^3$). The acceptable level for this known carcinogen in New York State is $0.13\mu g/m^3$. Their fears were confirmed. One of the many industries surrounding the neighborhood was the source, but which? From the wind direction, they suspected that Tonawanda Coke was the culprit.

"We weren't activists or anything," James-Creedon tells me over the telephone. "We were winging it." With evidence from their single sample in hand, they tried to talk to JD Crane, the owner of Tonawanda Coke, but were repeatedly refused meetings. They notified the US Environmental Protection Agency (EPA), but the agency had found Tonawanda Coke to be in compliance with emission standards during every inspection they *scheduled*. The physical law of inertia states that action does not come without applying force, and the same seems to hold true for regulatory protection. James-Creedon and her crew became forceful advocates for the health of everyone in Tonawanda. Over the next ten years, they helped guide a positive shift in the way regulators like the EPA view community monitoring.

A key turning point occurred when they found an ally at a state agency. Al Carlacci is the regional air pollutant control engineer with the New York Department of Environmental Conservation, and when he saw the benzene results he wanted to help. James-Creedon calls him "their champion." There are about twenty industries neighboring Tonawanda, and Carlacci knew it was essential to identify which was the source beyond any reasonable doubt. Science, and citizen science, in regulatory context has to pass the highest level of scrutiny, far beyond what's common for peer-reviewed papers. Carlacci applied for EPA funding and was awarded $600,000. He set up four state-of-the-art continuous air-quality monitors, each many times more expensive than the James-Creedon's simple bucket sampler. The time-stamped readings, along with measurements of wind direction, allowed Carlacci to not only further quantify the indisputably sickening amounts of benzene in the air but to triangulate and pinpoint the source, which was, as suspected, Tonawanda Coke. Carlacci's study motivated the EPA to pay an unscheduled visit to the facility, because their repeated scheduled inspections had not found problems. This new inspection revealed that Tonawanda Coke was intentionally violating emission standards and, confirmed by a whistleblower, hiding its actions during planned inspections.

In March 2014, a twelve-person federal jury in Buffalo, New York, found Mark Kamholz guilty for poisoning the air of Tonawanda with hundreds of tons of benzene-laden gas. The sixty-five-year-old environmental control manager for Tonawanda Coke was a serial polluter, releasing ninety-one tons of benzene into the air per year, creating cancer risk seventy-five times greater than allowable by federal law and emissions standards set by the EPA. He was found guilty of one count of obstruction of justice, eleven counts of violating the Clean Air Act, and three counts of violating the Resource Conservation and Recovery Act. His sentence: one year in prison, followed by one hundred hours of community service and a $20,000 fine. Tonawanda Coke is paying bigger fines, though the details are lingering in court battles.

This type of justice—in which a person in industry is held accountable for harming public health—is slow and scarce, and not particularly severe. The Kamholz verdict was only the second time in the United States that a corporate employee was convicted in criminal (rather than civil) court for polluting, and the first time such a conviction resulted in jail time. When Kamholz was sentenced, the US attorney for the Western District of New York, William Hochul Jr., said, "Today's sentencing holds Tonawanda Coke and its environmental control manager accountable for one of the most egregious environmental pollution crimes in this area's history." Those are strong words given that Tonawanda is only ten miles from Love Canal, the blue-collar neighborhood in the city of Niagara Falls where pollution was so intense in the late 1970s that the federal government ended up fencing off the neighborhood and evacuating it, relocating families elsewhere.[90]

There are small groups across the world facing similar struggles. When the neighbors in Tonawanda wanted to learn how to build a bucket-style air sampler, they reached out to Global Community Monitor, an organization that teaches communities how to make and use inexpensive tools to monitor their environment so that data can be part of the battles for environmental justice. Data gives people leverage to address problems, whether they are pushing for detailed follow-up studies or for relocation or cleanup. Community monitoring will continue to be important because thousands of hazardous chemicals come from industrial sources daily, and the EPA—and their equivalents in other countries—can't monitor them all. Global Community Monitor has worked with over a hundred partners in twenty-seven countries. Their air-monitoring system, known as the Bucket Brigade, has been replicated around the world, and people modify it according to local industrial sources and cultures. Pollution is not released with labels like the goods we purchase from stores; it remains unidentified until people find tools to reveal it. Global Community Monitor continually receives urgent calls from communities who want and need to uncover what hazards they may be exposed to.

As we've seen in earlier chapters, citizen science typically involves

---

[90] The town of Love Canal had failed economically by 1910, leaving the unfinished canal (essentially a ditch) as a receptacle for industrial chemical waste. In the early 1950s, the chemical dump site was eventually capped by industry and sold for one dollar to the city. The city grew, and by the 1970s, toxic puddles surfaced, vegetation died, and residents experienced a high number of miscarriages and birth defects. After continued Superfund cleanup (burying the toxins), Love Canal was declared safe in 1988, houses were renovated, and the neighborhood was renamed Black Creek Village.

hundreds, thousands, and even tens of thousands of people working toward a common goal. Crowds may be typical in citizen science, but they are not essential. James-Creedon quotes Margaret Mead: "Never doubt that a small group of thoughtful, committed citizens can change the world; indeed, it's the only thing that ever has." Global Community Monitor recommends a minimum core of five people monitoring air quality. Tonawanda had four, and they collected just one sample that triggered a cascade of investigations and ultimately a conviction and cleaner air.

A primary function of regulatory agencies is to establish whether federal or state air-quality standards have been violated, and they rarely accept volunteer data as valid for this purpose. According to Gwen Ottinger, an assistant professor of science and technology studies at Drexel University, the common approach is exactly what happened in Tonawanda: the use of the volunteer data to spark an in-depth— and expensive—professional investigation. According to Ottinger, when agencies follow up on community findings, they usually find that the pollution is worse than detected by low-cost devices like the air-sampling buckets. But not always. Sometimes a low-cost method simply reveals a known problem. For example, a community might detect an emission violation but it corresponds to a refinery reporting an accidental fire that day. Industry mistakes that are rare, self-reported, and fixed don't require further action on the part of a regulator. On the other hand, when the bucket sample reveals a problem that does not coincide with any industry reports, then it's time for some sleuthing. In Tonawanda, that's why the in-depth study triggered an unscheduled inspection from the EPA.

When James-Creedon and her neighbors in Tonawanda collected their air sample, they were guided by their noses and stinging throats. Given that air samples from citizen scientists are limited in their use to raising red flags and directing regulatory authorities to do a better job, the Louisiana Bucket Brigade (LABB), an organization with the stated mission of ending petrochemical pollution in Louisiana, created a way for anyone detecting pollution with any of their five senses—even if lacking a bucket sampler—to easily inform regulatory efforts. After all, when petrochemical pollution is in the air people can smell it, taste it, and feel it. LABB created a way for people to map it online.

I met Anne Rolfes, the founding director of LABB, at an EPA workshop on air-quality monitoring. Rolfes has led the fight against petrochemical pollution in Louisiana for fifteen years; she grew up in Louisiana, and her friends and neighbors worked for oil companies. As a young adult, she helped shine a spotlight on Nigerian refugees in Benin who were there because their farmlands had been destroyed by Shell Oil. (She subsequently wrote a powerful report titled *Shell Shocked Refugees.*) The plight of the Nigerians made her realize that the problems of the oil industry were universal, and that she should return to Louisiana, where people were also facing harm from an unchecked industry.

LABB's community-based air monitoring has long involved bucket samplers like the one used in Tonawanda, and has led to significant drops in industrial accidents and releases of toxic chemicals. But there is still a long way to go to ensure cleaner air. Rolfes explains that LABB's activism arises from the framework of a broken system: "An ordinary person's expectation would be that citizen scientists would present data to the responsible authority and then that authority would enforce laws and regulations and corporations would clean up their messes and stop making new messes. Unfortunately, that's not how the system works."

The area has over 130 chemical factories, petroleum plants, and hazardous waste incinerators. According to Barbara Allen, author of *Uneasy Alchemy,*[91] in 2003 the chemical industry in Louisiana reported the equivalent of sixteen thousand pounds of hazardous waste for every person in the state. Louisiana is renowned for corrupt politicians, and the petrochemical industry has a checkered

---

[91] Allen's book recounts how a pharmacist in St. Gabriel Parish noticed an unusually high rate of miscarriages about twenty years ago, which led to uncovering the effects of petrochemical pollution.

past. Put the two together along a river the size of the Mississippi (that draws those in the petrochemical industry who believe "dilution is the solution to pollution") and you get a river corridor now known as Cancer Alley, or Allée du Cancer to many of the French-speaking locals.

Rolfes and Iris Brown take about three hundred people per year on a biking and driving tour of Cancer Alley called Down by the River. Brown began fighting for environmental justice after losing family members to illnesses related to chemical exposure. Rolfes and Brown don't frame the tour as doom and gloom, but as activism and citizen science across boundaries of race and class. According to the LABB website, "Down by the River is an evolution of what has traditionally been called a Toxic Tour in which polluting sites—dumps, refineries, and chemical plants—are toured. The Down by the River ride transforms a litany of depressing sites into powerful beacons of resistance." Here the civil rights movement of the 1960s and the environmental movement of the 1970s intersected to create the environmental justice movement in the 1980s and the online crowdsourcing of the twenty-first century.

One of those beacons along Cancer Alley is the small town of Norco, in St. Charles Parish, west of New Orleans. In Norco, where Brown was born and raised, sits a Shell Chemical plant that frequently exposed residents to pollution. There was regulation intended to minimize pollution, but the financial penalties imposed by regulatory agencies were far lower than the profits Shell Chemical made by polluting. A primary way for the Shell plant to lower pollution involved lowering production, but that caused lower profits. Instead, Shell preferred to maintain high production even when it meant that excesses had to be burned in sudden flares that released toxic gases. Shell Chemical created a siren system to alert their neighbors of these incidents of acute pollution, and instructed residents that the siren meant to go indoors, close all doors and windows, seal off ventilation, and remain there until they heard a subsequent all-clear signal.

Residents formed the Concerned Citizens of Norco to fight for an alternative.[92] They campaigned for a long time to bring attention to the pollution, but they didn't have any leverage until they started using buckets to sample the air.

According to Rolfes, "Industry would rather have a community siren system to warn residents of a highly polluting flare than to temporarily lower their production and reduce pollution." Norco residents collected bucket samples as part of their campaign for compensation to relocate to homes elsewhere; these samples held Shell accountable when it misrepresented the frequency and size of accidental and intentional releases. Data on pollution actually made the people on the receiving end more visible to industry and got them a seat at the negotiating table. Brown played a big role with the Concerned Citizens of Norco and in persuading Shell Chemical to buy out tainted property so that residents could relocate to towns with clean air.

Rolfes is grateful for the bucket samplers and the relocation of Norco residents, "but Norco was just one battle, and it needs to be repeated a thousand times over. In the long term, the environmental justice fight will be won by changing the paradigm of our acceptance of major polluters like the oil industry. We need to get off this stuff all together. We just need to get rid of this stuff. Citizen science helps demonstrate the harm of petrochemical industries and offers a counternarrative; otherwise the narrative is one-sided and people only see the industry and government perspective. There are two sides, and citizen science helps us show our side. Otherwise the problems remain invisible." Rolfes subsequently began exploring other forms of citizen science that could be effective in giving voice to more people who bear the burdens of industry pollution.

LABB created an online map to crowdsource testimony about pollution. It partnered with Ushahidi,[93] an organization creating open-

---

[92] The Norco story was published in a 2003 study *Community environmental policing: Assessing new strategies of public participation in environmental regulation* by Dara O'Rourke and Gregg Macey, social scientists from the Massachusetts Institute of Technology.

[93] Ushahidi means "testimony" in Swahili.

source tools for crowdsourcing—particularly tools for mobile devices for use in emergencies. The Ushahidi platform was initially created to map reports of postelection violence in Kenya in 2008. Ushahidi hosts many projects, including QuakeMap, which has helped guide rescue efforts in areas hit by earthquakes, such as Kathmandu, Nepal, in 2015.

LABB calls their map the iWitness Pollution Map, and uses it to help focus the efforts of those putting out monitoring equipment. The crowdsourcing effort is based on the assumption that in Louisiana the EPA has been unable to adequately detect the extent of pollution and exactly who is exposed to it in quantities that put them at risk.[94] Years prior to the iWitness Pollution Map, the EPA put out air monitors and reported that these monitoring stations detected no major problems with air quality. When LABB analyzed the crowdsourced locations of testimonials reporting high pollution on the iWitness Pollution Map, it showed that the EPA monitoring sites were poorly chosen: they were nowhere near where residents were reporting odors. Rolfes explains, "We didn't have to say a lot. The maps visually showed that the EPA isn't monitoring in relevant places."

The process of creating the iWitness Pollution Map coincided with the 2010 BP oil leak in the Gulf of Mexico. The crisis prompted LABB to modify its plans in order to quickly launch the site first as the Oil Spill Crisis Map. An online map for crowdsourcing information about the disaster, it was the first use of crisis mapping for humanitarian purposes in the United States. In 2013, in continued partnership with Ushahidi, LABB morphed the Oil Spill Crisis map into the iWitness Pollution Map, which is now the world's largest site for community-gathered data on environmental pollution. On the Down by the River tour, tourists can add their observations (that is, sensory experiences) to the iWitness Pollution Map, just as those on bird tours enter their observations to eBird (see chapter 2).

---

[94] The EPA regulates six common air pollutants, which they call criteria pollutants: ground-level ozone, particulate matter, carbon monoxide, nitrogen oxides, sulfur dioxide, and lead.

The online mapping and the bucket brigades represent an array of organizing tools for communities to put pressure on industry to change. When I ask Rolfes if it would be possible for the petrochemical industry to not pollute, she responds, "We are so far from asking that question. We are dealing with problems that are so egregious like pipes with duct tape, leaking pipes, unlawful emissions, and ignoring basic engineering practices. It can be like the Wild West with unenforced laws." The common chemical leaks, flaring, and accidents are symptoms of a larger problem. Even while these symptoms need attention from citizen science monitoring, those seeking environmental justice can't ignore the underlying problem: a culture that protects industry instead of people.

People affected by industrial pollution are white and black, rural and urban, in the United States and across the globe. The only thing they generally have in common is that they are not wealthy. In many cases, there isn't a single fall guy like Kamholz to shoulder all the blame. Instead, there is a long-standing power structure of politicians and academics that intentionally or unwittingly shields industry from responsibility. Communities cannot easily overcome these power dynamics. What does citizen science look like in this context? Even when the source of pollution is different, the story is remarkably similar.

Throughout eastern North Carolina the hogs don't live on *Charlotte's Web* farms but in efficient industrial complexes that look like metallic military barracks. In towns near industrial hog farms, when the wind changes direction, children waiting in line in the school cafeteria begin to vomit from the putrid sulfuric stench of urine and feces; before stepping off the school bus in the afternoon, they cover their noses and mouths, then run into their homes with their eyes stinging. Families can't have guests over for dinner because they are embarrassed by the foul air that engulfs their home. Neighbors can't even imagine a cookout together, and no one hangs clean laundry outside. Elderly people in homes without air conditioning find opening the windows out of the question, even in the heat of summer. People who live next to these hog farms are struck with physical ailments as hog sewage literally rained down on them as it was sprayed onto adjacent fields to keep the sewage pits from overflowing.

In addition to the inequality problems that arise with being poor and black in the south, eastern North Carolina's problem with air pollution started with Wendell Murphy, who served as a Democrat in the North Carolina House of Representatives and later the State Senate; in his elected positions, he facilitated legislation that supported the growth of the swine industry in North Carolina. This type of hog farming resembles factory chicken farming: that means five thousand pigs in a building, each confined to a stall, with food piped in and large fans to ventilate. Each 250-pound pig produces fifteen pounds of manure daily. Collectively, that can be seventy-five thousand pounds of feces, plus untold gallons of urine, daily—all of which falls through slats in the floor and is flushed out and into unlined, uncovered ponds about thirty feet deep and up to eight acres in size. One hog farm can produce volumes of waste equivalent to that of a medium-size city. Yet a medium-size city would be a preferable neighbor, because municipal waste is treated in sewage plants while hog waste is not. The unlined sewage ponds and the bacteria in the waste pose a potential contamination risk to nearby private wells because the water table is high—the wells are only about fifteen to twenty feet deep. As the ponds fill up, the facility workers spray sewage onto surrounding fields. The wind carries a fine misty rain of urine and feces into neighboring towns, turning people into prisoners in their own homes.

In this system, soil, air, groundwater, and streams are at risk of contamination from bacteria, nutrient overloads, and even pharmaceutical drugs like antibiotics that are heavily used to stem infections among the overcrowded hogs. There are more pigs in North Carolina than people, and they are concentrated in a way that puts people at grave risk. Pigs do not make good neighbors.

The industry spins hog farms differently, as is explained to me by Naeema Muhammed, a community organizer and the director of the

North Carolina Environmental Justice Network, and Gary Grant, the charismatic leader of the Concerned Citizens of Tillery (CCT) in Tillery, North Carolina. The mission of the CCT is to promote social justice and self-determination for rural African American communities. Grant tells me that at a town meeting industrial lackeys told Tillery residents that they should feel a special honor to have hog breeds from England on their neighboring farms, as if excrement from the queen's subjects smells sweeter. He holds an imaginary tea cup with his pinky at a right angle, "Oh, and do we get tea and crumpets at ten o'clock?"

Grant and Muhammed try to one-up each other in stories about the hog industry representatives, politicians, and academics who act as if rural blacks are stupid. They have been told the sewage ponds were called lagoons, as though that should conjure images of beautiful girls splashing in the shallow blue waters of the South Pacific. They have been told the lagoons are not filled with sewage, but with "organic fertilizer." Muhammed recounts a time when a professor from North Carolina State University tried to convince Tillery residents that the waste from hog farms would be "no worse than herds of deer running through the forest." She adds, "When have you seen five thousand deer running through the forest, and all stop to poop at the same time, in the same place?" She and Grant joke, but they have been deeply insulted. Muhammed organizes such dedicated people in citizen science and activism, and it wears her down to see them incessantly being treated like they are dumb, spoken to like children, and never invited to the table for discussions.

Those who persist against such negative forces have one key trait that makes them resilient: a creative sense of humor. The North Carolina Environmental Justice Network (NCEJN) once received a permit to demonstrate on the capitol lawn. They set up inflatable kiddie pools with a sprinkler fountain and filled them by trucking in organic "lagoon" fertilizer from a nearby hog farm. To avoid the demonstration site, politicians were abruptly detouring from the usual path between their offices and the legislative chamber. The facilities manager threatened NCEJN: "If you spill one drop, we have to fine you because it is toxic waste." Muhammed recalls saying, "So you mean to tell me that we loaded a truck with organic fertilizer and by driving it forty miles down the road, it became toxic waste? Hmmph. Besides, the permit doesn't specify that we *can't* spray hog waste." The following year they got another permit to demonstrate. This time the permit specified in bold print, "No permission for hog waste."

Whether demonstrations, phone calls, or letters, ultimately every imaginable way that residents turned to elected government officials for help received the same basic response: "We were told to prove it," Muhammed explains. "The government put the burden of proof on the public to show that farms cause health impacts instead of telling the industry to prove it was safe."

Muhammed and I meet at the Curing House, which sits between the Tillery Community Center and the Tillery History House.[95] Don't let this cluster of three buildings mislead you into concluding that Tillery is a thriving town, because these are the only three buildings you can shake a stick at along a country road cutting through pine plantations about an hour east of Raleigh. The Curing House is an almost-defunct health clinic. There are empty spots where computers used to be when the building held examination rooms in which patients met online with doctors at the Brody School of Medicine at East Carolina University. Funds ran out, and now one doctor drives out and spends three hours each month seeing patients. There have never been doctors inquiring about illness related to the hog farms; instead, each doctor has had a "tell me what's wrong and I'll treat it with pharmaceuticals" approach. Yet Tillery is a data-driven community, and locals want to collect information on health patterns and the environment and fix the cause, not treat the symptoms. "Is that why you call this the Curing House?" I asked Muhammed. "No," she laughed, "when Tillery was still a community of farmers, this building is where the crops of sweet and white potatoes were put to cure."

---

[95] Students in documentary studies at Duke University helped the CCT design the History House.

The hog industry had been spreading across eastern North Carolina for decades before it got close to Tillery. Hog cities were built and their sewage pits dug adjacent to rural communities that are predominantly black, with high unemployment and poverty. The industry leaders assume citizens in those communities lack political clout, don't know anything or anyone in the state capital in Raleigh, and have no champions in politics because they don't have spare change to donate to politicians. Plus, their land is cheap. Robert D. Ballard, author of *Dumping in Dixie*, called such communities "avenues of least resistance."

Based on demographics, tax records, and the highest number of homes with outdoor plumbing in any county in North Carolina, anyone might mistake Tillery for an avenue of least resistance, as the hog farming industry did. To the contrary, Tillery is a vibrant and organized community, and when it got wind of plans for over fifteen facilities in the area, the townspeople rolled up their sleeves for a fight. The burden of proof foisted on them by elected officials was one of many burdens they already carried, and they set out to gather evidence to set the record straight.

Because of its fairly unique history, Tillery might have been different from otherwise similar towns that were unable to resist the encroachment of hog farming. Tillery was established as a farming community as part of President Franklin Delano Roosevelt's New Deal in the 1930s; by establishing the Resettlement Administration in 1935, Roosevelt hoped to end tenant farming and sharecropping. The program aimed high—hoping to help over 500,000 farm families own farms—but ended with only 4,441 families relocated to land they could call their own. The situation was reminiscent of Reconstruction after the Civil War, when former slaves and descendants of slaves were given forty acres of land and a mule, only to have it soon taken back in a way that led to sharecropping becoming the dominant agricultural system by the 1870s. A poster advertising the Resettlement Administration even included a drawing of a black farmer with a mule and the words "A mule and a plow." A second wave of black settlers with the dream of farming came to Tillery in 1947 with Farmer's Home Administration loans.

When Grant gives me a tour of Tillery's History House, we watch a film titled *We Shall Not Be Moved* that includes interviews of residents of Tillery. Maggie Crowell, never looking up at the camera, recalls her youth seventy years earlier: "I come up in fear because I thought my mommy and daddy would get hurt." Grant explained that Crowell was looking away from the camera because she was afraid to look at the white camera crew. The deeps scars of Jim Crow laws never fully heal.

Of the older residents in Tillery and surrounding areas, Grant says some have always been activists, continuing to fight all forms of oppression, and others just want to leave whites and their industries alone in hopes of being left alone themselves. Either way, people travel far across rural North Carolina to attend CCT meetings; Grant says they come because it is a liberating experience. For example, one day before a meeting, Crowell was at the Tillery Community Center when a white representative from the Methodist Church dropped in to verify whether CCT was a legitimate organization because it had applied for a grant. The representative confronted Grant: "Why have *I* not heard of CCT before?" Grant retrieved and proudly displayed an enormous stack of news clippings about CCT and asked, "Now you tell me why you have never heard of CCT." Crowell was elated by the exchange, and later said to Grant, "You can talk back to white folks like that? I didn't know. Times *have* changed!"

Grant invites me to lunch at the Tillery Community Center with a group called the Open Minded Seniors to meet some community members. I learn that many of these senior citizens are raising their grandchildren, and sometimes their great-grandchildren, while their own children have moved away to take jobs, many in Baltimore, to support their families. As is universally acceptable among grandparents, they proudly boast of the intelligence, good grades, and accomplishments of their grandchildren. Although these seniors "shall not be moved," they are sad that the next generation will undoubtedly move away for jobs elsewhere. Industries have tried to come to Tillery with promises of jobs, but offers like a battery recycling plant, a facility to turn New

York City's sewage sludge into fertilizer, and any offers you can imagine that are low on wages and high on pollution have been easy for Tillery residents to reject. These were farming families; "forty acres and a mule" had seemed like a good deal, but even that one was broken by systematic exclusion from farm aid that should have come from the US Department of Agriculture.[96]

After the main meal and before dessert, Grant takes a picture of three people in their nineties. Instead of prompting them with "Say *cheese*" he uses "Say *sex*," which elicits big smiles. Everyone who has a birthday that month sits at a special table, and we sing "Happy Birthday" to them all, naming each person in turn. Grant announces the age of each; one woman writes down their ages and, like an awesome science nerd, reports that their average age is 76.3.

Grant concludes the gathering with a joke about pollution.

A police officer asks a man why he is putting garbage in a ditch. "Can't you see the sign right above?"

"Yes, sir. It says, 'Fine for putting garbage in the ditch.'"

Grant identifies as "second-generation New Deal," meaning his parents arrived in the second wave of settlers. He was a hippie in the 1960s, engaged in the civil rights movement, and boasts nineteen civil disobedience arrests. He still calls people like me "white folks," but qualifies it with "don't take it personally." A pivotal time for Grant was when North Carolina governor Jim Hunt was going to allow PCBs to be dumped in Warren County. During protests of this, Grant heard the term *environmental racism* and it instantly made sense; he transitioned from the civil rights movement to the environmental justice movement. Like that representative from the Methodist Church, whites in the Sierra Club, Environmental Defense Fund, or Greenpeace sometimes

---

[96] In 1997, a class-action suit led by Timothy Pigford sued the USDA and its secretary Dan Glickman for racial discrimination in the delivery of farm aid programs throughout the 1980s and early 1990s. In 1999, the plaintiffs won $1.4 billion in damages. In 2009–10, a second phase of $1.2 billion was signed into action by President Barack Obama. It was a bigger moral victory than an economic one: after dividing up the sum among all those affected, each farmer was awarded write-offs on any loans and about $50,000 as taxable income.

wonder, Why do *I* not know any black environmentalists? African American communities and other disenfranchised minorities experience an unprecedented burden of environmental pollution and struggle against power imbalances to restore environmental health, while white environmentalists are donating money to save the whales. Until traditional environmental groups also include environmental justice in their mission and vision, the environmental movement will remain highly segregated.

It is in the Curing House after lunch that Muhammed explains to me that citizen science fit into their environmental justice strategy in two ways. One way was to examine the geographic distribution of the hog industry in relation to the geographic distribution of people of color and thereby expose whether environmental racism was present; the other way was to gather time-synchronized data on levels of air pollution from hog farms and the physical health of residents in order to see if the relationship showed evidence that farms were physically harmful for residents. For both purposes, Muhammed and Grant carried out citizen science in collaboration with Steve Wing, a professor at the School of Public Health at the University of North Carolina–Chapel Hill.

To expose environmental racism, Muhammed and Grant worked as citizen scientists with Wing and his students to create maps, which were also a community organizing tool. By overlaying maps of race and income characteristic of each county and seeing where farms were and were not located, they calculated that industrial hog farms were more likely to be located next to communities of the poor and people of color than near a town with wealthy white residents. They published the results in a scientific journal. The only thing worse than bearing the brunt of pollution is being saddled with an unfair amount of it.

To document whether farms were negatively impacting public health, they gathered data. Muhammed recruited and trained residents in sixteen communities near hog farms to monitor their own health and keep records. After sitting outside twice a day for ten minutes, residents filled out data sheets about their immediate health (such as watery eyes, headache, irritated throat) and used small devices to time-stamp and automatically measure their blood pressure, gather a saliva sample (used to measure stress hormones), and quantify their lung capacity. In tandem, UNC professor Wing had deployed small trailers full of automated equipment to measure concentrations of particulates and chemicals in the air over time. Was public health correlated with pollution levels?

A common argument leveraged by science skeptics and Internet trolls is that correlation does not mean causation. For example, increases in spending on science may correlate with increases in per capita consumption of cheddar cheese, but the two variables are completely unrelated: unfortunately, eating more cheese won't increase my likelihood of getting science funding. Nevertheless, correlation is one necessary, but not sufficient, condition of finding a causative relationship, and that's why the statistical process of correlation is integral to research. Another condition to establishing causation is having a set of prior reasons (a priori hypotheses) about the mechanism of the causation. In this case the expected mechanism was hydrogen sulfide, which was known to be a hazard in laboratory studies. Did the amounts in the air in the real world matter? Short answer: yes. What made the study with Wing rigorous was that his automated devices measured ups and downs in hydrogen sulfides in the air over time as it fluctuated with wind and manure-spraying schedules. Thus, if airborne hydrogen sulfides in the real world caused health impacts, then the personal health data would fluctuate up and down according to the exact same temporal schedule; if hydrogen sulfides *did not* cause health impacts, then health data would not correlate with hydrogen sulfides in the air at all.

When Wing combined data about air quality with the time-stamped data residents had collected on their personal health, the correlations were obvious enough to establish causation. Wing published several papers, some with Muhammed and Grant, showing that personal health declined when hydrogen sulfide levels in the air spiked, and health increased when these levels were low. *Snap!* They had proof to take to the state legislature.

Muhammed tells me, "We call it community-based participatory research." In fact, the term, often referred to more simply as CBPR, has been used longer than the term *citizen science* and describes a process in which community members and professional scientists work together to carry out science to gain knowledge useful to the community. CBPR puts the community in the driver's seat while the scientist navigate from the passenger seat. Muhammed rattles off a list of reasons why communities do it. "The benefits include increased understanding of the scope of justice, raising awareness, educating policy makers, informing legal disputes, and contributing to the knowledge base. Most importantly, it adds legitimacy." Teachers and other educators focus on science, technology, engineering, and math (STEM) learning from citizen science, particularly among populations underrepresented in the STEM workforce. When I ask Muhammed about STEM she says, "Sure, we learned to read charts, graphs, and report results back to the community." Even though the primary motivation for studying hogs and health was not to motivate Tillery youth into science careers, STEM learning is just as essential to community organizing as it is to recruiting students into the science career pipeline.

People of color in the Deep South have health issues that arise from multiple factors, but race and geography also matter. African Americans in the South are more likely to have heart disease, cancer, diabetes, AIDS, and health markers that indicate higher vulnerability to ill health, such as high blood pressure and obesity. The communities least able to handle additional health risks from hog farms are the ones most likely to get it.

The struggles continue, though the activists won one battle when the state legislature passed a moratorium that prevents new hog farms from the spraying of sewage from the open "lagoons." As the industry began losing, the fight got ugly. The hog industry targeted Wing as the community's most powerful accomplice. Academia is structured around rewards, like peer-reviewed publications, that keep science busy within the ivory tower. Academic interests usually don't conflict with industry, and sometimes even align, if science can help bring home the bacon. This was not the case with Wing, who had given people access to the validated system of science and with it the ability to discover authoritative knowledge to take to the legislature, and that ran counter to industry interests. By not following traditional academic incentives, it was as though he breached an unspoken code. The Pork Council decided to put him in line, lest other academics follow suit; the council took Wing to court to force him to breach a clearly articulated ethical code among researchers: confidentiality. It obtained court orders directing Wing to hand over the citizen science data set created with the help of Mohammed and Grant. It would be an unethical breach of confidentiality to reveal the names of the people who participated in gathering health data about themselves. After a long court battle, UNC lawyers could not help Wing protect the identities of the citizen scientists, but Wing would not be intimidated, so he hired his own lawyer with his own money. The solution he crafted was to agree to hand over the data set *after* anonymizing it— that is, stripping all unique information that would allow individuals to be identified.

Scientists like Wing, willing to incur industry ire, place priority on responding to the needs of communities. Community and industry interests don't always conflict. But in situations where the industry causes the air to have a foul smell, taste, and sting, or where it has to put a siren system in place, or where it literally causes feces to rain down on people, identifying the problem should be a no-brainer. Instead rigorous citizen science and lots of public pressure are need to change industry practices. In these cases, scientists need to decide whose interests they will serve.

It isn't just academics in the field of public health who respond to community interests. Let's detour to Flint, Michigan, where government, rather than industry, decisions led community members to take up citizen science in efforts to balance the scales of environmental justice. The problem in Flint began when unelected officials, appointed as emergency managers to run the city, decided to alleviate the financial crisis by switching water supplies. The switch, which occurred in April 2014, disconnected the city from the Detroit Water and Sewage Department, which drew water from Lake Huron, and connected it with the Karegnondi Water Authority (KWA), which drew water from the Flint River. By June, Flint resident LeeAnne Walters noticed that she had a rash. This might not have been a cause for concern had it not occurred in conjunction with her husband, her older son JD, her older daughter Kaylie, and her toddler twins Gavin and Garret also developing rashes, hair loss, and a host of other, unexplainable symptoms of illness. The Flint Water Study, the citizen science project developed to address the underlying problem, unofficially began a year later when Walters telephoned Marc Edwards, an engineering professor at Virginia Tech.

At that point, the Walters family had already stopped using Flint tap water. But for eight months Walters' children, like many in her community, had been exposed to the water—and the lead it contained—and suffered irreversible harms. By the time she called Edwards, Walters was certain that the water was the source of their health problems. Others in her community knew it too, but everyone they turned for help—city, state, or federal authorities in various agencies—refused to own up to the problem.

After spending months visiting doctors and slowly progressing through a series of unnecessary treatments based on incorrect diagnoses, the situation in the Walters' household intensified at the end of 2014. Just after Christmas, brown water started coming through the Walters' faucets with increasing frequency. In January 2015, Walters received official notice that there was a safety violation related to their water. The violation noted excessive levels of a carcinogenic substance used to disinfect water—not lead. At a City Council meeting on January 7, 2015, staff distributed out-of-date handouts, evaded questions, and simply assured residents that everything was fine.

"We had been struggling with illnesses and had believed it was just our home, our plumbing, that had a problem. At the first City Council meeting, we realized it was the whole community," Walters explained to me over the phone. To address public concerns, Flint officials held a town hall meeting on January 21, 2015. "Everything felt wrong about the meeting: the location was changed at the last minute to a smaller venue, no one was sworn in, and police presence was apparent." Not only were no solutions in sight, but no one seemed willing to acknowledge a serious problem. On the drive home, Walters and her family pondered a key question: "How do we get the authorities to hear us? We decided we had to figure out the science of it because the authorities won't be able to argue with science." For people like Walters, attaining the authority of science in order to sway political authorities is a common approach when needing to protect themselves from hazards in the environment.

Walters started reading everything she could find about water management as well as laws related to water. She came to understand the federal Safe Drinking Water Act, and a key EPA regulation called the Lead and Copper Rule, inside and out and every loophole these contained. She explained, "I began gathering information and trying to figure out the bigger picture." By the beginning of April, Walters discovered that the City was not using a corrosion inhibitor in the water. "I was puzzled and thought maybe polymer aids, which were present in the water, could be used for corrosion control. But when I asked Miguel Del Toral[97] about polymer aids, he said no, they couldn't provide corrosion control. Together we realized that the city was breaking federal law by not using corrosion control!" This discovery coincided with one of her twin sons finally being tested for lead and receiving an official diagnosis of lead poisoning.[98] Walters' extensive research led her to Marc Edwards, whom she called to back

---

[97] Del Toral was the EPA regulations manager and expert on lead.
[98] *Plumbum* is the Latin word for lead and the basis for the English word "plumbing." Some speculate that plumbism (lead poisoning) led to the fall of the Roman Empire. In those days, people knew lead was dangerous in heavy doses of acute exposure, but failed to realize that low-level daily exposures caused infertility, stillbirths and miscarriages, and low intelligence. Today, we know of the dangers of lead and that children are the most vulnerable to harm from exposure to it.

her findings and solicit help for the next stages. Walters noted how easy it was to discuss her findings with Edwards: "Scientific people can sometimes make you feel stupid or beneath them, but it was never that way with Marc. He is blue collar. He is one of us. He explains things in ways that are not condescending."

My own conversation with Edwards affirms Walters' assessment; Edwards told me, "Even though I'm a professor, I'm basically a plumber at heart who just wants to help people. Even though I don't go around unclogging toilets every day, all I do is study plumbing systems and how to improve them to protect public health."

Through a follow-up phone call, Edwards guided Walters step-by-step through an intense sampling procedure to obtain thirty bottles of tap water. Walters immediately sent the samples to Virginia Tech. Edwards recalled, "About a week later, the results showed that LeeAnne's water contained literally hazardous waste levels of lead. One gulp could raise the lead level of a child to over the CDC level of concern. Thank goodness they weren't listening to the state's claims that the water was safe."

"First we wanted the EPA to do their job," Edwards explained. Allies in the EPA like Del Toral agreed. He wrote a memo that outlined the dangers in Flint and requested that the EPA exercise emergency powers to intervene. Unfortunately, those higher up the chain of command did not value Del Toral's assessment of the situation. He told Edwards that an EPA ethics officer told him to never again speak to anyone from Flint, or about Flint.

When Del Toral's memo failed, residents were left in imment danger. Edwards launched the Flint Water Study in mid-2015 as a citizen science project in which residents could follow a simple protocol to collect tap water in their homes, ship it to Virginia Tech, and receive results about water chemistry, particularly lead. He also felt the site would document an environmental crime. Flint residents needed access to scientific methods, just like residents in Tillery, Tonawanda, and New Orleans. Edwards said, "We provided the funding, technical and analytical support that allowed Flint residents to save themselves."

Even with data in hand, Walters and other residents still had to fight to be heard. Authorities tried to discredit the citizen science. "Fortunately, Dr. Mona's blood work of kids made it that much harder to ignore," Walters said, referring to Dr. Mona Hanna-Attisha, who led a study that retrospectively examined blood lead levels in young children before and after the change in Flint's water source. For many years, the Hurley Medical Center had routinely tested the blood lead levels of children, creating a database with information about lead levels from over one thousand children living in Flint and over 2,000 living outside of Flint. Hanna-Attisha found that outside the city, blood lead levels did not change, but within the city of Flint, the incidence of high lead levels increased after the change in water supply.

Ultimately, President Obama declared Flint as a federal state of emergency and authorized help from the Federal Emergency Management Agency and the Department of Homeland Security. Michigan's Governor Snyder allocated millions to aid Flint residents and replace lead pipes. The Michigan Attorney General has been filing criminal charges against an assortment of people for crimes including tampering with evidence, misconduct in office, and willful neglect of duty. Walters continues to campaign to close every loophole in the EPA's Lead and Copper Rule so that others won't fall victim to lead poisoning.

This wasn't the first time Edwards had listened to people suspicious about problems with their water. He knew from past experiences, including finding lead in water in Washington, DC, that when someone complains about their water, they are often correct in noticing a real problem.

Professional societies for scientists, like the American Association for the Advancement of Science, and science funding agencies, like the National Science Foundation, are encouraging scientists to hone their communication skills. With public skepticism about global climate change, genetically modified foods, and vaccinations, these organizations view science communication as a responsibility of scientists and a necessity for informing the public on scientific issues. Unfortunately, most scientists communicate as though giving a lecture in class; they talk *at* people rather than converse *with* people. Edwards and his colleagues at Virginia Tech excel as science communicators in the opposite way: they *listen* to people.

In fact, Edwards and Yanna Lambrinidou, in the Department of Science and Technology Studies, take the skill of listening so seriously that they offer a college course in which students learn to listen to the public. The course is called Engineering Ethics and the Public, and the focus is on what they call transformational listening. By honing listening skills, researchers can collect information—not by measuring or observing but by listening—to inform their research agendas to meet public needs. At the same time, the conversations challenge stereotypes, expose power inequalities in the relationship, and transform relationships into trusted partnerships.

Another listener is Mordecai ("Muki") Haklay, professor at University College London, who studies a broad range of topics related to environmental inequities driven by geography, particularly using citizen science approaches in response to the interests of local communities. Haklay began the Extreme Citizen Science lab, which collaborates with nonliterate communities around the world, and he cofounded Mapping for Change, a community mapping platform to help communities add their voice and their data to counter environmental injustices in London. Haklay's approach to citizen science is at the grassroots level, while most scientists pursue topics that are trending in the literature or stated priorities of funding agencies. When I ask Haklay how he decides what topics to study, he replies, "Meh," humbly waving my question aside, "It's not rocket science. I go into a neighborhood, sit in a café, and have conversations with people."

Edwards, Wing, and Haklay are examples of a new breed of STEM professionals who are skilled in shaping their research agendas to serve public interests. (We'll meet another, Julia Brody, in chapter 10.) Unfortunately, these types of scientists are still few and far between. More often, when communities look for academic allies, they tend to find only those dedicated to their own research agendas. Given the importance of citizen science to justice, and the need for scientists to support citizen science endeavors, it is high time to re-vamp public expectations.

Public expectations are related to perceptions of scientists, which are typically reflected in movies and literature: middle-aged or elderly white men in glasses and lab coats carrying out dangerous experiments, like Dr. Frankenstein, or with dual personalities like Dr. Jekyll and Mr. Hyde, or bumbling nerds like Sherman Klump (played by Eddie Murphy) and Julius Kelp (played by Jerry Lewis) in each of the very different movies called *The Nutty Professor*. For decades, common perceptions were that scientists aimed to do tremendous social good, but sometimes their curiosity got the better of them and they were so absorbed by their work that they possessed few social skills.

The stereotypes are reinforced by television, movies, and comic books, and never countered because most people do not personally know a scientist nor recognize one even after the Science Cheerleaders (see chapter 6) point them out. Now that citizen science is putting more people in touch with scientists, and more scientists are defying stereotypical behaviors, could firsthand contact flip both public perceptions and expectations of scientists?

The answer could affect the diversity of people who imagine themselves as scientists and are happy with science careers. Although governments, academia, and foundations have repeatedly sought to increase the number of underrepresented minorities in STEM careers, progress has been slow, even stagnant. Andrew Campbell, a biology professor at Brown University, carried out candid discussion with underrepresented minorities (URMs) considering STEM careers in order to understand the pertinent issues. Campbell's synthesis pointed to impediments related to the culture of science, which wasn't teaching students about career paths soon enough, nor providing solutions for work-life balance, nor—most important—valuing diverse traits and skills through different metrics of success. Campbell and his colleagues reported that to stay interested in STEM careers, URMs want scientific work to connect with communities and intersect with the humanities and arts. Science in silos shuts out too many types of people. Campbell recommended adding a social justice component to STEM education, as well as training in science communication so that students could help their families understand what research careers were all about. Expecting students to learn to collaborate with communities via citizen science designs seems like another approach that might broaden pursuit of STEM careers.

In chapter 10 we'll explore highly varied forms of citizen science in public health, a field that values both social and environmental justice.

CHAPTER 10

Public Health
Patients, No Patience

I'm no longer accepting the things I cannot change. I'm changing the things I cannot accept.

—ANGELA DAVIS

ONE OF THE WORLD'S FIRST INKLING OF THE DISEASE THAT WOULD EVENtually become known as AIDS was a headline in *The New York Times* on July 3, 1981: "Rare Cancer Seen in 41 Homosexuals."[99] Before the end of that year, AIDS would be recognized as an epidemic. By 1987, there was only one AIDS drug with federal approval, azidothymidine (AZT), and activists in the Gay Men's Health Crisis, which formed in 1982, wanted more options; they wanted a cure. By that point, more than forty-six thousand Americans were infected with HIV, the virus that leads to AIDS, and many of the thirteen thousand already deceased were being honored in panels on the AIDS Memorial Quilt. In 1989, the quilt was nominated for the Nobel Peace Prize; today it contains over 48,000 panels, each a colorful three-by-six foot memorial. As some people mourned and pieced together the quilt, others pieced together what was needed to discover effective treatments quickly.

Starting in the late 1980s and continuing to the present, AIDS activists have organized numerous demonstrations to speed the process of drug approval and transform medical science to make it responsive to the AIDS crisis. In 1988 it was Seize Control of the FDA, organized by ACTUP (the AIDS Coalition to Unleash Power): splattering blood at the doorstep of the US Food and Drug Administration (FDA), activists demanded rapid

---

[99] The "rare cancer" mentioned in the article was Kaposi's sarcoma, an opportunistic infection commonly associated with AIDS.

approval of experimental drugs to treat AIDS and the ability to try treatments from other countries because AZT and other drugs were not effective enough. Next ACTUP turned its attention to the National Institutes of Health (NIH). On May 21, 1990, for example, about a thousand activists attended a demonstration called Storm the NIH. Instead of a sit-in or teach-in, the standard peaceful choices for activists in the 1970s, they staged a die-in, lying on the ground as if dead, some next to mock tombstones. ACTUP held die-ins not just at the NIH but also at numerous pharmaceutical company headquarters. Today when the pharmaceutical industry spends over $27 billion per year advertising and pedaling new drugs to people,[100] it's difficult to imagine a time when the public found it necessary to demand pharmaceutical companies make more types of drugs. At the push of activists, the NIH and big pharma made new drugs, and the FDA changed its regulations so that drugs were evaluated more quickly. Activists reduced the time frame for testing the safety and efficacy of AIDS drugs so they could count down time until FDA approval in months rather than years.

It was the development of the HIV antibody test in 1985 that propelled some activists into science. Anyone could take the HIV antibody test and learn they were infected before showing any symptoms, yet still, four years into the epidemic, there were no effective treatments and a positive test result was a death sentence. Since the gay community was most affected, and was already overflowing with seasoned activists, it organized efficiently to search for a cure, working with many types of scientists: immunologists, virologists, molecular biologists, epidemiologists, physicians, and biochemists. These activists realized that there was one primary goal in seeking help through these credentialed scientists, the only goal that mattered against a ticking clock: speed. Treatment delayed was treatment denied.

AIDS activists fundamentally reformed the clinical trials of AIDS research in the late 1980s and early 1990s. They became credible agents

---

[100] According to a special report by the Pew Charitable Trusts in 2012, the pharmaceutical industry spent over $24 billion marketing drugs to doctors and over $3 billion marketing drugs to consumers.

within the scientific community, viewed as partners to AIDS researchers. These activists reveal an entirely different idea of what citizen science can mean. So far, we've used the term *citizen science* to describe various ways laypeople collect or analyze data to advance research; the term is also used more broadly to describe ways that laypeople participate in and influence the practice of science. Indeed, the term was initially coined by Alan Irwin in 1995 to mean just that. This type of citizen science requires scientists who are receptive to public needs and concerns—"public scientists," as Rob Dunn (chapter 6) calls them.

Irwin, a British sociologist, introduced citizen science as a way to describe a more democratic, participatory science. Science may seem to be carried out within institutions insulated from society, but it is deeply embedded in a social matrix. Irwin emphasizes that science is a human creation. Nevertheless, it is not perfectly situated in society, and certain aspects of science are driven by institutional forces like commercial markets, politics, and hot trends in academic publishing that do not necessarily represent the needs and desires of the general public. Irwin's work—published in a 1995 academic book titled *Citizen Science*—addressed this issue by analyzing two dimensions of the relationship of citizens with science. The first is that science is for the common good and should seek to address the needs and concerns of people. The second is that the process of producing reliable knowledge can be enhanced by laypeople. Irwin wrote, "People bring into science such things as local contextual knowledge and real-world geographic, political, and moral constraints generated outside of formal scientific institutions."

Let's look at an example of AIDS activists as citizen scientists. Because the activists weren't collecting data as we've seen with other citizen scientists, they fit into the concept of citizen science later described by Irwin. Consider Mark Harrington, a screenwriter with no scientific background. Harrington, an early member of ACTUP, took part in protests in order to draw public attention to the AIDS crisis, but he also sought to change scientific practices. He grabbed every resource he could—textbooks, journals, medical reports—and taught himself enough of the technical details of AIDS so that he could participate knowledgeably in scientific discussions. In January 1992, members of ACTUP's Treatment and Data Committee, including Harrington, left the group and founded Treatment Action Group (TAG), a nonprofit organization focused on accelerating AIDS treatment research.[101] That same year, Harrington delivered his first plenary address at the Eighth International AIDS Conference. He began coauthoring peer-reviewed papers, and has continued to publish to the present day.

How did Harrington go from street demonstrator to scientific collaborator? In an insightful dissertation (and eventual book, *Impure Science: AIDS, Activism, and the Politics of Knowledge*), Steven Epstein, now professor of sociology at Northwestern University, described exactly how AIDS activists like Harrington took a four-pronged strategy to gain credibility and authority.

First, AIDS activists learned to speak the language of researchers, entered the culture of medical science, and then transformed it. As with learning a foreign language by spending time in another country, immersion was best. Activists attended conferences, critiqued research papers, found tutors to help them understand cell biology. Harrington prepared a fifty-page dictionary of the vocabulary relevant to AIDS research. Once activists could speak the language of viral assays, reverse transcription, cytokine regulation, and epitope mapping, scientists were more receptive to engaging in discussions.

Second, activists presented themselves as the voices of people who were suffering from HIV/AIDS. They were the brokers, ensuring that researchers could enroll enough HIV-positive people in their treatment trails and helping the participants to understand and comply with experimental treatment protocols. People with HIV/AIDS needed researchers, but researchers needed their help and cooperation too. Thus, activists gained leverage when negotiating the finer points of clinical trials.

Third, early AIDS activists changed the way clinical trials were carried out. Early trials were limited to middle-class white men; that didn't make sense given that populations affected by AIDS also included drug

---

[101] To this day TAG remains the leading citizen science group in AIDS research internationally, and Harrington is the organization's longtime executive director.

users, hemophiliacs, women, and people of color. Federal (and international) laws requiring ethical oversight of clinical trials were designed to protect trial participants from harm, particularly those in vulnerable communities. The laws, which even today are highly paternalistic, were predicated on myriad past abuses in medical research in the United States—in particular, a study of a different sexually transmitted disease.

In 1965, Peter Buxtun, a social worker and epidemiologist, began working for the US Public Health Service. I don't know what was printed on his business card, but his job was venereal disease investigator. Within a year he began calling for an end to the long-term Tuskegee Syphilis Experiment (officially called the Tuskegee Study of Untreated Syphilis in the Negro Male). In the Tuskegee Syphilis Experiment, which began in 1932, hundreds of black men with syphilis were deceived into believing they were receiving health care; instead, they were studied while doctors withheld medical treatment. Even after penicillin was found to cure syphilis in 1940s, after Buxton voiced concerns in 1966 and again in 1968, after William Carter Jenkins called for the experiment's end in 1968, and even as wives and children were becoming infected, the Communicable Disease Center insisted the study go to completion.[102] "Completion" in this case meant all study subjects had died and been autopsied. The Tuskegee study was no secret among scientists. They published research papers, and doctors reading scientific journals knew exactly what was happening, but the public did not. The only way to end the study was to let the public know. In 1972, Buxtun "leaked" details to the press, and public outrage triggered congressional hearings. As a consequence of this egregious violation of the Nuremberg Code, the Declaration of Helsinki and, of course, the Hippocratic Oath,[103] the US government now mandates ethical oversight of research with human subjects.

---

[102] Today the Communicable Disease Center is known as the Centers for Disease Control and Prevention.

[103] The Nuremberg Code and Declaration of Helsinki were created in response to atrocities by medical researchers experimenting with concentration camp detainees during World War II.

The laws do protect human subjects, but quite often, individuals and communities under study have learned that they also need to protect themselves. Self-determination in research has given rise to the mantra of community-based public health research, a slogan that embodies citizen science philosophy: "Nothing about us, without us." Ironically, the nature of legally required institutional oversight can inadvertently undermine the spirit of the law. For example, human subjects of research are "protected" by not having ongoing contact with researchers, and by not receiving information about their personal data. Those kinds of rules may protect passive volunteers, but they restrict the autonomy of engaged participants. In citizen science, an ongoing relationship between scientists and community members is essential, as is the sharing of data collected.

AIDS activists were handed clinical trials carried out by a system that was overcorrecting for historic abuses. But this was not Tuskegee, Alabama, anymore. Instead, activists were championing the idea that experimental treatments were a social good to which everybody should have equal access. They convinced researchers that people have the right and autonomy to assume the risks of experimental therapies and be informed partners in research.

Fourth, activists improved AIDS research by influencing the design, conduct, interpretation, and speed of clinical trials. Researchers used to carry out only randomized, controlled, clinical trials, and many of the controls limited access to potentially helpful treatments. Why should any patient with a terminal illness be given a placebo, and why should a patient who had already tried one treatment be disqualified from trying another? Researchers wanted clean data from highly controlled experiments. But the real world is messy. Activists convinced researchers that drugs should be tested in real-world situations with heterogeneous groups. Not only was it fair, but the answers would be quick and more reliable. The only way to obtain clean data in a messy world was to unfairly manipulate and control people. By emphasizing that AIDS clinical trials were simultaneously research and medical care, activists knocked years off the time frame for testing the safety and efficacy of AIDS drugs.

The configuration of citizen science participation is almost as varied as the ways proteins can fold. On one side of the coin, citizen scientists play video games and donate their spare computational power to speed the process of finding effective ways to treat a range of diseases; on the other side of the same coin, citizen scientists reform research to make it responsive to their needs.

AIDS activists were the type of citizen scientists described by Irwin. They gained the type of authority that usually comes from academic degrees, and they went from diseased victims to activist experts. Today such activists serve on institutional review boards of hospitals and research centers. They are representatives at FDA advisory committee meetings where drugs are considered for approval, and they are voting members of NIH committees that oversee drug development.

AIDS activists are one extreme example on the spectrum of the citizen science participation. After all, most diseases don't strike preexisting interest groups. Gay men and lesbians were well positioned to become AIDS activists; they were already politically organized thanks to the gay liberation movement of the 1970s. They were already pursuing civil rights, and they had already "demedicalized" gayness. They had resources, people of influence, funding, a strong public relations arm, lobby groups, and community organizations. This was what strong social networks looked like before the Internet.

But with the Internet, even more connections are possible. Many diseases create disabilities that prevent people from physically gathering, but via the Internet they can form social networks to advance research on their disease. Citizen science is filling the prescription for social networks among individuals managing chronic illnesses. This dose of citizen science is called patient-led research.

Online social networks can help medical research on rare disease. For example, there is a rare cancer, gastrointestinal stromal tumor (GIST); its propensity to metastasize rapidly makes it very deadly. Because it occurs in fewer than fifteen out of every million people, researchers have difficulty getting a sufficient number of patients together for clinical drug trials. In 2000 the drug company Novartis initiated a clinical trial for its new drug

Gleevec, for which it needed eight hundred patients. An Internet-based nongovernmental organization, the Life Raft Group, collaborated and gathered data from GIST patients around the world taking the drug— dosage, side effects, response to treatment, and even mortality rates. Although it was not a randomized trial, it was a larger sample than had ever previously been possible. This was citizen science on human subjects, where people contributed data about themselves. In this case, the data showed that those taking a low dose of Gleevec died, and those taking high doses lived longer.

Another example is amyotrophic lateral sclerosis (ALS; also knowns as Lou Gehrig's disease), for which Stephen Hawkins (the one scientist people are most likely to have heard of, according to polls) is an anomaly. Hawkins was diagnosed with ALS when he was twenty-one years old, and has surpassed all odds by living as long as he has (he is now in his mid-seventies). Tens of thousands of people, across all racial and ethnic groups, are diagnosed with ALS annually and their fate is, more typically, death within five years. PatientsLikeMe (PLM), a sort of Facebook for disease research, was started in 2004 by two brothers, Jamie and Ben Heywood, and their friend Jeff Cole. The Heywoods' brother, Stephen, had fallen to ALS. PLM enables social networking for the sharing of medical experiences, data, and insights. It is sustainable as a for-profit company because medical data, as well as access to patients, are commodities to medical and pharmaceutical researchers. PLM is citizen science because, as we've seen before, it is necessary to share data, to bring minds and experiences together—particularly when time is of the essence.

PLM embraces the principle that getting data into more hands will speed the pace of medical research and improve the health care system. The PLM business model is to have pharmaceutical companies and medical researchers as customers because these groups are willing to pay for access to a network of patients interested in advancing research on treatments. Over 350,000 people share information on over two thousand illnesses in the PatientsLikeMe network, and patients learn from other patients. They learn to navigate their relationships with health care providers, and they help pharmaceutical companies bring treatments to market in record time.

People never lie down and die without a fight, and in the age of the Internet, they do not have to fight alone. In 2008, clinicians in Italy presented research results at a prestigious health care conference, showing data indicating that lithium delayed the progression of ALS. Their sample size was small, only sixteen patients, so it could have been a fluke, but the report was intended to encourage further research by other clinicians.

An ALS patient in Brazil, Humberto Macedo, used Google translate to read the conference abstract. Word spread among ALS patients, and many wanted to try lithium treatment. They could get lithium from friends or sympathetic pharmacists because the drug is widely used to treat bipolar disorder. Macedo started a website to recruit ALS patients into a clinical trial with the drug—a clinical trial designed by participants.

When PatientsLikeMe saw the ALS patients organize for an experiment, it decided to modify its website to accommodate the patient-led clinical trial. PLM was not condoning or condemning the choices of those with a terminal illness to carry out an unsupervised medical experiment. Rather, given the fact that the patient-led Lithium trials were going to happen, it wanted to make sure the symptoms of the disease and side effects of treatment could be reported in a consistent manner with standardized scales because that makes for better data. The group wanted to make sure that the study would produce good science that would benefit other patients and researchers in the future.

In addition, PLM created an algorithm to match each patient self-administering the lithium treatment with three to five people who were opting to not self-experiment so that the experiment would have a control group. In a few months (which is a large fraction of the time ALS patients typically have left) there were 160 participants.

The cooperation put into the research was incredible and inspiring, but the outcomes of this citizen science experiment were not. Sadly, the patients did not get the same results as the Italian clinicians had.

Lithium didn't work to slow ALS, and it might have made things worse. ALS ran its course. When citizen scientists allow themselves to be the subject of the research, they let their bodies supply the data. Macedo and the others died making a valuable contribution to medical research.

Citizen scientists who are not terminally ill also participate in clinical trials within PLM. For example, another PLM community has formed around epilepsy, which affects over two million people in the United States. A traditional approach to help epilepsy patients manage their condition, which is frequently accompanied by depression, is for health care professionals to contact people by telephone via the Managing Epilepsy Well Network. But it turns out that patients benefit more from helping each other. Through PLM, those with epilepsy track their health and share the reporting of their symptoms and treatments with other patients just like them; they share their experiences and their data. PLM has created tools specific to the epilepsy community with sponsorship from pharmaceutical company UCB, and every epilepsy patient on PLM can see health records of every other patient who opts to share. Patients learn more about their own illness through this type of sharing, and they get support from others dealing with the same conditions. The result is improved health outcomes; for example, 55 percent of epileptics in the PLM community agreed they learned more about seizures, and 27 percent said PLM helped them stick to their medication. Almost 20 percent said that participation in PLM caused them to need fewer visits to the emergency room. Remarkably, almost 30 percent said participation in the PLM social network reduced the negative side effects of their medications.

Data about personal health that doctors collect, even during routine visits unrelated to research studies, are protected by two federal laws: the Health Information for Economic and Clinical Health Act and the Health Insurance Portability and Accountability Act (HIPAA). HIPAA does not extend to online social networks like PLM, but PLM abides by the spirit of HIPAA anyway by removing all names and information that would allow a person's identity to be revealed before selling data to the pharma, biotech, or insurance industries. When it comes to medical data, a high level of mutual trust and transparency is needed for people to share their information. Put individuals in control of their personal data so that they own it and can loan it, and then the benefits begin to outweigh the privacy risks.

In a 2014 survey of social media users, 94 percent were willing to share their personal health data anonymously with doctors in order to improve the care of others with similar problems. Almost as many, 92 percent, were willing to share their data with researchers. An overwhelming majority were willing to share with drug companies, with slight differences related to how the drug companies would use their data: 84 percent were willing to share in order to help drug companies make safer products, and 78 percent if it were to learn about diseases. At the same time, three-quarters of respondents suspected their health records could already be used without their knowledge. People were concerned about negative consequences, with 72 percent worrying that their records could be used to deny them health care benefits, 66 percent believing records could be used to deny them a job, and 61 percent worrying that their data could be stolen.

When researchers were solely in charge and patients were kept in the dark, there were heinous abuses in medical research. The consequence of those abuses is the paternalistic system of institutional oversight of professionals carrying out research with human subjects. How does oversight work if research is patients led? How can patient privacy be protected in the Internet age? How do the benefits of open data weigh against the privacy risks? The solutions to these ethical quandaries will find their way through citizen science practices. Ultimately, if you want to provide clues to the answers to medical questions, you may have to raise your hand and be counted.

Here's an example of the speed of research made possible by a repository of medical data. Sally Okun of PLM led a 2013 study about MetroHealth Medical Center in Cleveland, Ohio, and its partnership with an analytic company called Explorys. MetroHealth is a safety net hospital, meaning it provides care to the uninsured and people with low income, and it is affiliated with Case Western Reserve University School of Medicine. With MetroHealth, researchers compiled fourteen million medical records, from which Explorys could replicate a longitudinal Norwegian study of heart disease risk. A longitudinal study is one that follows select individuals over time, rather than a cross-section of the population at once. In Norway, researchers followed more than 26,000 individuals for thirteen years and found that the risk of blood clots was highest for men and linked to obesity and height. In the MetroHealth /Explorys study, researchers could see the same patterns within three months of sifting through the data because the information had already been collected incidentally as part of routine health care. Plus, they had almost one million relevant records, so their estimates of risk were more precise. The Norwegian study costs millions of dollars; the MetroHealth /Explorys study of "big data" had a price tag of only $25,000.

The field of public health, which is aimed at preventing disease and promoting human well-being through research, education, and organizing, is also breaking ground with citizen scientists who are not patients. Many don't want to steer research agendas but simply want to lend their time toward helping. They don't necessarily need, or want, to understand diseases, and they may even avoid understanding the details of diseases and their symptoms to avoid triggering hypochondriac tendencies. But with high stakes and no time to lose, thousands of people are analyzing data by making their way through online tasks, which leads to the faster development of new treatments.

The goal of leading cancer research charity Cancer Research UK is to ensure that 75 percent of patients survive cancer by 2035. It plans to do this by focusing on cancer prevention, diagnosis, treatment and through the optimization of cancer care. Personalized medicine, which aims to find the best possible treatment for each patient by examining the biological details of the patient and his or her cancer, is a key aspect of this strategy. Researchers do this by looking closely at the patient's genes, and the genetic makeup of the cancer cells.

There are over two hundred forms of cancer, and Cancer Research UK supports research into all of them, with a special focus on cancer types that are poorly studied and/or hard to treat—in the latter case, specifically lung, pancreatic, esophageal, and brain cancer. Around half of us are predicted to get cancer at some point in our lives, so it's a disease that has a very real impact on everyone. Public interest in cancer research was one reason why Cancer Research UK began tapping the potential of citizen science.

Cancer researchers were also in desperate need of citizen science to help them in their work. Increasingly, scientists are using technology to quickly process lots of samples and collect large amounts of data to help them better understand cancer and develop useful treatments. There are so much data around that there aren't enough scientists to keep up. In cancer pathology research, which involves studying samples of cancer cells under a microscope, there is a major backlog of samples in need of processing. Professional pathologists are busy full-time in clinics and hospitals to help diagnose cancer, leaving tens of thousands of samples intended for research that are prepared each year but wait for examination. Worryingly, the number of people training to be pathologists has fallen, which means that there are very few trained professionals available to study these samples. Given that there are millions of samples waiting to be examined, not enough pathologists to do the work, and computer algorithms not yet advanced enough to automate the analysis of these images, citizen science has been a great alternative.

For this emotionally heavy subject Cancer Research UK relies on the help of over 500,000 citizen scientists doing things like playing specially crafted video games to help understand the disease. At end of 2014, the research charity celebrated the equivalent of fifteen years of collective volunteer time that was squished into twelve months. By the time the program closed in March 2016, it had released four citizen science projects to the public, amassing over eleven million individual analyses from people in 182 different countries. It also showed that results of greater than 95 percent accuracy could be achieved through citizen science, proving that citizen science could be a viable option for the analysis of cancer genome and pathology data.

Cancer Research UK citizen science projects follow a similar model to Galaxy Zoo (see chapter 4) in that they have samples already collected by professional scientists and ask citizen scientists for help to study these samples.

One project, called Cell Slider, is hosted by Zooniverse (see chapter 4) and has engaged ninety-eight thousand participants in contributing close to two million analyses of over twelve thousand samples collected from around six thousand breast tumors. Cancer Research UK has stores of thousands of tumor samples that are dissected to create millions of photographs. Each sample came from a patient for which the associated researchers have information on treatment and outcome.

Each photograph shows blood cells, tissue cells, and cancer cells.[104] By drawing the help of online crowds, researchers were able to analyze data much faster than would typically be possible in the lab. Using treatment and outcome information from these former patients, together with the information from citizen scientists, allows researchers to achieve better foresight to improve treatments and personalized medicine for future patients.

While Cell Slider was a fairly standard online citizen science project that simply presented images and asked scientific questions, Cancer Research UK went on to look at the potential of building citizen science into computerized games. The charity's hope was to engage more people in citizen science for cancer research by making its projects fun and portable while giving players a feeling of progression and improvement as they played, thus encouraging them to keep playing.

In March 2013, Cancer Research UK sponsored a forty-eight-hour

---

[104] The cancer cells are bigger than the others and irregular in shape. If participants view a cluster of cancer cells, then they are asked to guess how many are in the slide, to estimate proportion of cancer cells that appear yellow relative to those that appear blue or pink, and to assess the brightness of the yellow color. A similar project focused on bladder tumors is called Reverse the Odds.

competitive hackathon, called GameJam, to develop its first game. It challenged forty competitive computer programmers and hackers, working in small teams, to embed raw, anonymized genetic data into a video game in which players would detect genetic faults within cancer cells during gameplay. The intent was to make the analysis fun, rewarding, and accessible to a wide range of people. The winning game is a mobile app called Genes in Space, designed by Guerilla Tea. Genes in Space is a first-person game where the player is a pilot who navigates a spaceship to collect a mysterious futuristic fuel source while being pelted with asteroids. The locations of the fuel are actually data from DNA microarrays indicating the number of copies of different genes. A normal cell should have two copies of each gene, but cancer cells can have fewer or many more. Notably, although the scientific detail behind Genes in Space is not fully explored in the game, most people who played it reported that the fact that it contributed to real scientific research was an important part of why they played. These sorts of games may be the beginning of a new trend intersecting citizen science and gaming.[105]

The impetus for Genes in Space, to increase the speed of analysis of data on gene expression in cancer, came in 2012 from the work of a team led by Carlos Caldas at Cancer Research UK's Cambridge Institute. Caldas published a study in *Nature* that found that breast cancer is not one but ten different diseases, each with its own molecular fingerprint, and each with different weak spots for targeted treatments.

As an example, over one hundred years ago, researchers found that some breast cancers form because their cells produce too much estrogen receptor (ER) protein, and this acts on cancer growth like sunshine on a seedling. For decades, doctors have administered antiestrogen drugs like tamoxifen to combat these specific types of breast cancer cells, but this only works well when the cancer has too much ER. To personalize this treatment, breast cancer cells for a patient are tested for their level of ER protein. Progesterone receptor (PR) protein is also in some breast

---

[105] Within a week of the release of the virtual reality game Pokémon Go, biologists were using the Twitter hashtag #PokeBlitz to encourage players to take photos of real wildlife while on their hunt for Pokémon characters.

cancer cells. Patients with cancer characterized by PR protein are more likely to respond positively to antiestrogen drugs; patients whose cancer cells contain both PR and ER proteins (denoted PR+ and ER+ and called double-positive cancers) have the best outlook with these drug treatments. This leaves women with only one positive (either ER+ or PR+) or with neither (called double-negative cancers) with fewer options and in need of more help.

Thanks to molecular biology, in the last two decades research has accelerated and found more variation in cancer types. In the late 1990s researchers discovered that breast cancer cells produced a protein called HER2, which responded to the drug trastuzumab. In personalized medicine, women are tested for ER, PR, and HER2 and treated accordingly. Women who are triple-negative currently have fewer effective drug treatment options. The variation, and the need for personalization, does not end there.

More recently, research on breast cancer switched from measuring ER, PR, HER2, and other *proteins* inside of tumors to looking at which *genes* are toggling cancer cell growth on and off. Now, instead of categorizing cancers by the proteins present (which are made by genes), breast cancers are categorized into genetic subtypes. These include luminal A cancers, luminal B cancers, HER2-amplified cancers, and basal-like cancers. Unfortunately, while understanding these variants is helpful for understanding the disease and choosing treatments, most variants don't currently have an effective personalized treatment. Researchers suspect there is a hidden underlying pattern. Caldas and his lab are part of the Molecular Taxonomy of Breast Cancer International Consortium (METABRIC), which aims to further differentiate between breast cancers to aid in identifying optimal treatments for all breast cancer patients.

The strategy of the METABRIC project is data intensive. From each patient, with permission, the project stores a sample of noncancer cells and a sample of tumor cells. It then records the treatment and eventual outcome of the patient associated with each sample (such as whether they became cancer free, spent years in remission, or died from the cancer). When METABRIC overlaid the gene sequences from tumor cells with sequences from healthy cells for each person, and did so repeatedly for thousands of people, they found local hot spots—small segments of genes where problems tend to arise. Each hot spot might map to certain genetic mutations specific to breast cancers. Researchers suspect it's quite possible that a single mutation affects lots of cell processes associated with cancer. To investigate further, Cancer Research UK turned to game-style citizen science.

Genes in Space was released in February 2014 and within a month achieved what would have taken scientists six months: players collectively performed over one million analyses, examining about forty miles of DNA for these potential hot spots. As people play Genes in Space, their moves piloting a space ship are turned into information to understand the activity of different genes within cancer cells. Just as an angry person can suppress that emotion and look perfectly unfazed, so can DNA have a mutation but not express it as cancer. Mutations in much of the genome will not have a cancer-causing effect, and many healthy human cells can contain tens of thousands of mutations at any one time. Examining many different cancer cases helps to identify which mutations are actually relevant to causing cancer; this information can then be used to more accurately test patients for cancer, reliably identify different cancer types, and provide insight into possible ways to develop treatments for these different cancer types.

As one of the leading causes of death in the most economically developed nations in the world, cancer has long been a major public, political, and scientific concern, but until now research into cancer has been exclusively performed by professional scientists and doctors. These projects mark an exciting shift in the study of cancer where virtually anyone can contribute directly to developing treatments and cures that will see millions more people surviving cancer for longer.

While the majority of disease research is aimed at finding cures, a smaller percentage is aimed at prevention. The words of Catholic archbishop Dom Hélder Câmara are relevant to the philosophy of pursuing the cause of cancer instead of the cure: "When I give food to the poor, they call me a saint. When I ask why the poor have no food, they call me a communist." So far we've seen the "saint" fight against cancer in the search for cures and treatments, and now we'll see the wrongly branded "communist" searching for underlying causes to prevent cancer in the first place. This is citizen science in environmental health (as we saw in chapter 9).

All traits, even cancers, are the result of some mixture of our genetics and our environment. Genetics may bring predisposition to certain cancers; for example, based on studies of identical twins, inherited genes explain just over a quarter of breast cancer risk. Similarly, environmental exposures account for varied amounts of risk too; for example, solar radiation increases the risk of skin melanoma—according to the Skin Cancer Foundation, a person's risk for skin melanoma doubles if the person has had five or more sunburns.

Rachel Carson's 1962 book *Silent Spring* sparked a grassroots environmental movement by raising awareness about the risks of using chemical pesticides. The book led to a ban on the pesticide DDT and propelled formation of the Environmental Protection Agency. Were she alive today, Carson (who started out as a marine biologist) would likely be writing about plastic pollution in the oceans, but two years after the publication of *Silent Spring*, Carson died fighting breast cancer, about which she wrote, "For those in whom cancer is already a hidden or a visible presence, efforts to find cures must of course continue. But for those not yet touched by the disease and certainly for the generations as yet unborn, prevention is the imperative need."

The Silent Spring Institute in Massachusetts is a research organization with the goal of preventing breast cancer; it pioneers research by engaging communities in searching for links between environmental contaminants and cancer. In a phone conversation, the institute's director, Julia Brody, explains to me that its founders were influenced by AIDS activists and the type of citizen science they exemplify. They weren't necessarily interested in collecting data to help scientific research but instead in influencing research agendas. The founders were primarily residents of Cape Cod who discovered that breast cancer rates were higher for women on the Cape. Predominant cancer organizations, then and now, have focused on the search for a cure rather than take on the issue of preventing breast cancer. Brody explains that the institute is "a new model of citizen science where an expert science team is governed by laypeople." Expanding the idea of public scientists further, the research agendas of Brody and her colleagues are governed entirely by public interests, not the interests of corporations, or funders, or even hot topics driven by big-name researchers. "We are," Brody emphasizes, "not just grassroots, but citizen-owned-and-operated science." The founders wanted to balance research dollars devoted to a cure with dollars devoted to uncovering the cause.

The first step to preventing breast cancer is to find the environmental cause. The first place to look was at the high breast cancer rates on Cape Cod. With just over 200,000 year-round residents, the cape has an incidence of breast cancer that is 20 percent higher than the rest of Massachusetts, controlling for age and other factors like family history, reproductive history, use of pharmaceutical hormones, use of alcohol, and diet.

Over the last fifty years, hazardous chemicals have become ubiquitous—at work, in consumer products, in building materials, and even in our oceans and drinking water. Most chemicals have never been tested to determine what level of exposure is too dangerous. There are three classes of dangers that Brody focuses on: mutagenic chemicals (which cause mutations), endocrine disruptors (which disrupt natural hormones, such as chemicals that mimic estrogen), and developmental toxicants (exposure to which in youth increases a person's risk of cancer in adulthood).

Of those that have been tested to be confirmed mutagens, 216 cause mammary gland tumors in rodents (none are tested in humans). Of these, about one hundred are common: seventy-three in consumer products or contaminants in food, thirty-five in air pollution, twenty-five associated with workplace conditions. Twenty-nine of these are produced in the United States and forty-seven are pharmaceuticals. For example, mutagenic chemicals include benzene, which is in gasoline; ethylene oxide, used in food processing; methylene chloride, used as an industrial solvent; and many pesticides.

Those that are endocrine disruptors include bisphenol A (BPA), which could be found in the plastic in baby bottles, sports water bottles, water supply pipes, and food storage containers;[106] certain chemicals in personal care products like cosmetics and lotions, chemicals found in laundry detergents and other household cleaners; and—again— chemicals found in pesticides. After testing the urine from a representative sample of the population six years and older, the Centers for Disease Control and Prevention concluded that virtually all of us (93 percent) have BPA in our bodies. Another widespread cause of endocrine disruption are brominated flame retardants, which are commonly used on upholstery and carpeting and can find their way into human blood and drinking water.

The class of chemicals that are developmental toxicants include dioxin,[107] the pesticide atrazine,[108] and diethylstilbestrol (DES), a synthetic nonsteroidal form of estrogen;[109] if fetuses are exposed to any of these in utero, they have increased cancer risk in adulthood.

According to Brody, our society supports industry and the economy by granting an "innocent until proven guilty" approach to chemical

---

[106] BPA-based materials are no longer used in baby bottles, sippy cups, and infant formula packaging. BPA is the subject of much research, and it is still used in water bottles and to line steel food cans; many companies have stopped using it, and they typically label their goods as BPA-free.

[107] Dioxins are chemical by-products of industry processes such as smelting, chlorine bleaching of paper pulp, and the manufacture of some herbicides and pesticides. Most people consume dioxins in their food.

[108] Atrazine is a popular herbicide in agriculture and lawn care in the United States, and is commonly detected in drinking water. It was banned in the European Union in 2004 because of groundwater contamination. Some disputed research found atrazine exposure to cause the development of hermaphroditic frogs.

[109] Between 1938 and 1971, millions of women in the United States took DES to prevent miscarriage. In 1971, researchers realized that women given DES during pregnancy had a 30 percent higher risk of later getting breast cancer, and their daughters—exposed in utero—also have a higher risk of breast, vaginal, and cervical cancers.

additives. How's this approach working out for us? According to the National Cancer Institute (a division of the National Institutes of Health), cancer is the leading cause of death worldwide. Odds are that about 40 percent of people in the United States will be diagnosed with cancer at some point in their lifetime. The "innocent until proven guilty" mind-set, which is great for criminal justice, is counterproductive and misleading in medicine and in media reports. When reports say "there is no evidence that underarm deodorants and antiperspirants cause breast cancer," what's omitted from the statement is that "no evidence" could mean there has not been a proper study yet because of lack of funding or because of methodological obstacles. Brody gives this as an example of alternative wording: "The effect on breast cancer risk of using underarm products has not been investigated in a study that carefully compared women who use these products to women who do not. It is hard to study the effects of products that are widely used, because researchers cannot identify enough unexposed US women for comparison."

Lab studies with animals show that hundreds of common chemicals can cause mammary gland tumors in rodents. The tumors arise because the chemicals cause mutations in DNA and/or act as hormone disruptors. Hormone disruptors cause tumors to proliferate and, in young rodents, cause the mammary glands to develop in a way that makes the rodents more susceptible to cancer later in life.

We know from controlled laboratory experiments that hundreds of chemicals cause cancer in rodents, and we know that people are exposed to these chemicals. Yet few studies have been able to establish cause and effect in humans. The complexity of the real world is too great an obstacle. Exposures occur over years and are rarely measured because we can't *see* these chemicals. The latency from time of exposure to developing cancer can be years or even decades. People move from place to place, making it hard to track environmental exposures.

Pinpointing the causes of cancer is perhaps most straightforward with occupational hazards. In 1775, for instance, a London doctor name Percivall Pott figured out the connection between chimney sweeping and rampant scrotal cancer: our bodies turn the chemicals in soot into BPDE, a mutagen. Mary Poppins's Cockney friend Bert and other chimney sweeps should not *chim-chim-cheroo* in loose shirts and baggy trousers. As a result of Pott's discovery, chimney sweeps took to either bathing daily or wearing superhero-tight suits to keep soot off their skin.

Christopher Wild, the director of the International Agency for Research on Cancer, coined the term *exposome* in 2005 to refer to the sum of personal environmental exposures over a lifetime. Any trait that a person has, including diseases, are a combination of nature and nurture or, in this lingo, a product of one's genome and exposome. Participants in Brody's research share information not only about their health but also about where they've lived (and when) and details about household products. Brody and her team use that information to understand each individual's exposome.

Does the high occurrence of breast cancer on Cape Cod indicated that the cape environment causes cancer? No. With contagious diseases like the flu, epidemiologists can track the locations of occurrences and deduce the source of the cause. Breast cancer is not contagious, and also not necessarily contracted immediately or detected immediately. It is a developmental disease, which means that causal exposure could occur decades before it occurs and is diagnosed. This time lag makes it almost impossible to track the cause. Susceptibility to breast cancer begins in the womb and other key periods of life, such as puberty and menopause, are also susceptible times. To understand why there is high breast cancer rates on Cape Cod, Brody began by reconstructing chemical exposure histories for each volunteer in the study. Her team also examined concentrations of known carcinogens like DDT that persist in the environment. Brody found that two-thirds of the homes on the cape contain DDT, and at higher levels than in other cities.

One early piece of research was the Household Exposure Study, where the Silent Spring Institute team collected air, dust, and urine samples from homes on Cape Cod to see what was present. No one had measured the occurrence of endocrine disruptors indoors at that point. The Household Exposure Study was different from mainstream research agendas because it was informed by the grassroots start of the institute. Brody describes the research as "very connected science. We go into people's houses. We are there for an hour vacuuming for dust samples and collecting air samples. We come back later for their urine sample." Obviously the research requires an engaged community and close partnership between scientists and participants. This type of citizen science is not simply about people helping scientists collect data but about scientists helping people solve mysteries.

The collaborative relationship between scientists and people in the Household Exposure Study led to sustained efforts that called for new perspectives in the field of research ethics. At the time, the tradition was to not report back exposures to chemicals unless those exposures were known to be clinically relevant. "But," Brody explains, "our study was the first to measure exposure to various chemical compounds indoors, so no one knew what was typical or safe. In that situation, the traditional response was to not report." In other words, the traditional responses was no response. In clinical trials, the ethical behavior is for scientists and human subjects of the research to interact as little as possible and for human subjects to not receive information about their own data. But with this community-engaged model that is no longer the ethical response. Brody is concerned about how to ethically translate research to the public and patients to support informed decision making even while the science remains uncertain. She and her colleagues lead the way in guiding ethical approaches to citizen science with so many unknowns and so much at stake.

The next step for the Household Exposure Study involved a collaboration with an environmental justice group in Richmond, California, called Communities for a Better Environment. Brody's team trained local citizen science leaders there in the protocol used on Cape Cod for collecting household samples and gave them the equipment to do it. "This is another model of citizen science," notes Brody, "where a community-based environmental justice group is trained to carry out the research." The study was not simply repeated verbatim. When designing the study together, Communities for a Better Environment and the Silent Spring Institute expanded the list of chemicals for testing because local concerns focused on the neighboring Chevron petrochemical refinery;[110] the team was thus responsive to what people wanted to know. The indoor air in homes in Richmond was more polluted than the air outside. The outdoor pollutants from the oil refinery penetrated the houses, and this added to the burden of pollutants from consumer products. Levels of PBDE in dust in California homes were much higher than in Cape Cod homes. The State of California has strict flammability standards for foam used as padding under the upholstery of sofas and armchairs, and this caused manufacturers to add far too much flame retardants to their products for sale in California. Consequently, Richmond homes have a sum total of high pollutant exposure.

The Household Exposure Study was influential. The results formed the basis for knowing that consumer products are a major source of exposure to endocrine disruptors. Companies were claiming that chemicals remain in these products and are not released into the environment, but the research in these households showed that the industry claim was false. Thanks to these findings, in January 2015 California revised its standards for foam flammability, allowing manufacturers to add lesser amounts of flame retardants.

At the end of the day, Brody and her colleagues have not found strong links of cause and effect for breast cancer in the real world. That's because it is a near impossible thing to do: finding patterns in such a complex system is extremely difficult, like solving an algebra equation with too many variables. In 1976, President Gerald R. Ford signed into law the Toxic Substances Control Act, which was intended to protect people from chemicals that cause cancer and birth defects. When the law went into effect there were already sixty-two thousand chemicals in everyday products, and regulation is too difficult. Since then the US Environmental Protection Agency has banned only five

---

[110] The community ended up using the data to win a court case that required Chevron to do an environmental impact assessment as part of its planned expansion.

chemicals. Even the ban on asbestos hasn't stuck. Brody argues for a new paradigm: guilty until proven innocent. If lab studies show a strong carcinogenic effect or endocrine disruption, then that is reasonable doubt and the chemical should not be allowed in consumer products until it is proven to be safe. With developmental diseases like cancer, we should not expose people to potential problems that can take decades to come to light. In 2016 President Barack Obama signed into law the Frank R. Lautenberg Chemical Safety for the 21st Century Act. It ends the catch-22 created by the Toxic Substances Control Act and no longer requires the government to have evidence that a chemical poses a risk before it can require industry to test that chemical. Now industry has to test all chemicals before they enter the marketplace.

We've seen citizen scientists take charge of the research process, contribute to crowdsourcing, personalize their medicine, and establish community-owned and -operated science, all in the name of managing their own health. When people are ill, the custom is for them to passively receive information from health care providers whom they view as authorities and to seek a cure to treat their illnesses. When we buy products, we expect regulatory authorities to be looking out for our well-being and keep us safe from invisible hazards. Citizen scientists, whether empowered patients or activists, seek, evaluate, and synthesize information collaboratively with their health care providers, researchers, or community- based organizations and take charge of monitoring their health and their surrounding environment. They live well despite their illness or exposures, and investigate questions about the underlying causes.

According to Kelly Moore, a sociologist at Loyola University Chicago, citizen science can either undermine or reinforce the political authority of science. If citizen scientists are only providing unpaid work, then they are reinforcing the existing structures of the scientific enterprise. Although "unpaid work" and toeing a status quo line may sound unimpressive, this type of citizen science can result in revolutionary science: it can lead to amazing discoveries, as we've seen throughout the chapters in this book. But revolutionary science is different from a revolution in the structure of science. Some of the types of citizen science that we've seen in this chapter do change the power structures of science in ways that aid social justice. When citizen scientists are the initiators or play a large role in setting the research agenda, alongside scientists who listen and collaborate, then they are challenging power structures. The concepts of empowerment and democratization are complex in the context of scientific discovery. Gwen Ottinger, an assistant professor of science and technology studies at Drexel University, makes the case that for citizen science to fulfill its transformative potential, it has to deal with disparities in wealth, education, and power. These institutionalized imbalances play a role in environmental justice cases where there are often competing claims about knowledge and who has the expertise to produce it.

Science does not have to be an elite activity available to the small fraction of the population with graduate degrees. Science does not belong to any one group. In 2015, Effy Vayena, an ethicist, and John Tasioulas, a philosopher, laid out the case for citizen science as a fundamental human right. They interpret Article 27 of the United Nations 1948 Universal Declaration of Human Rights to mean everyone has right to actively participate in the scientific enterprise:

(1) Everyone has the right to freely participate in the cultural life of the community, to enjoy the arts and to share in scientific advancement and its benefits.

(2) Everyone has the right to the protection of the moral and material interests resulting from any scientific, literary or artistic production of which he [sic] is the author.

In addition, Article 15(1)(b) of the UN's *Report of the Special Rapporteur in the field of cultural rights* underscores that "*access must be to science as a whole*, not only to specific scientific outcomes or applications" (emphasis added).

Vayena and Tasioulas argue that if there is a human right to science as a whole, and not just its products, the state or other agents have the obligation to promote citizen science. After all, it is widely accepted that people have rights to take an active role in politics and culture, so why not in science too? We've seen citizen science enable people to follow their curiosity, contribute to a meaningful endeavor, learn about science, improve their health, and strengthen community action. Now let's see citizen science transform science into a tool for justice, which is the biggest reason to consider citizen science a human right.

We tend to hope for heroes, even superheroes to rescue us. We hope the next Albert Einstein will have the intellect, the next Mother Theresa will have the compassion. Citizen science can remind us of the collective power of people. We don't need to wait for a hero—we need to compel ourselves to greatness. Together we make new knowledge that scientists cannot make alone. Together we leverage social capital to create just and sustainable solutions. How should we navigate our future together? Let's conclude with a look at how we navigate our oceans.

CONCLUSION

Setting Sail

We are caught in an inescapable network of mutuality, tied in
a single garment of destiny. Whatever affects one directly, affects all indirectly.

—Martin Luther King Jr.

ACCORDING TO HOMER'S EPIC POEM *THE ODYSSEY*, AFTER DESTROYING
the ancient city of Troy around 1188 BCE, Odysseus begins a
decade-long journey home by sea. After many trials, tribulations,
and near misses, Odysseus loses his entire crew to Poseidon's raging
sea. In due course he is ferried home by skilled mariners, the Phaeacians.

For centuries to come, seafarers faced risk of shipwreck because they
navigated without full knowledge of impeding storms and without pen-
etrating the mysteries of currents, winds, and water depth. They blamed
the enchanted song of the sirens, attributed problems to arguments with
moody gods, and imagined backroom deals for bags of wind. Feeling
helpless and at the whim of the high seas, sailors became superstitious,
believing in bad omens ranging from sharks to bananas. The sailors'
braids, beards, and long nails were to avoid jinxes associated with
grooming. Their table manners, or lack thereof—such as stirring tea
with a knife or fork, but not a spoon—were to avoid bad luck. If they
were cruising at thirteen knots, deckhands reported it as 12 + 1. Instead
of bridging gaps with sound knowledge, seamen were gambling across
every void.

To make headway in lessening the dangers of seafaring, the best
maneuver turned out to be for sailors to document and share their
observations to collectively unlock the secrets of the seven seas. In
the 1840s, Matthew Fontaine Maury united mariners in scientific
pursuits and the perils of ocean travel retreated in the face of im-

proved understanding of the deep sea. Citizen science launched the discipline of oceanography and made ocean travel safer, faster, and more efficient.

It's worth taking a closer look at Maury and his approach to citizen science, particularly in comparison to William Whewell, who we met in this book's introduction. Superficially the two figures appear as similar as two white men in the mid-1800s asking ordinary people in faraway places to give them very particular observations. Yet there are fundamental differences in their approaches to science—and citizen science—that are relevant to how we, scientists and nonscientists alike, decide to structure our joint engagement in scientific inquiry.

Maury was a career navy man, having enlisted in 1825 when he was nineteen years old. His first assignment was as a midshipman on the frigate USS *Brandywine*. He worked his way up the ranks and ended his career as superintendent of the US Naval Observatory.

Today we'd call Maury a public scientist and science writer. His lyrical prose captured the attention of a broad audience. Early in his career, he became the first US naval officer to write a book on nautical science, which later became a textbook for the US Naval Academy. "The spirit of literary improvement has been awakened among the officers of our gallant Navy. We are pleased to see that science is gaining voteries from its ranks," praised Edgar Allan Poe.

Maury wrote of viewing a star in 1841,

At the dead hour of the night, . . . I turn to the Ephemeris and find there, by calculations made years ago, that when that clock tells a certain hour, a star which I never saw will be in the field of the telescope for a moment, flit through and then disappear. The instrument is set; the moment approaches and is intently awaited—I look—the star mute with eloquence that gathers sublimity from the silence of the night, comes smiling and dancing into the field, and at the instant predicted even to the fraction of a second, it makes its transit and is gone. With emotions too deep for the organs of speech, the heart swells out with unutterable anthems; we then see that there is harmony in the heavens above; and though we cannot hear, we feel the "music of the spheres."

Before astronomy had Carl Sagan, Ann Druyan, and Neil deGrasse Tyson, it had Matthew Maury.

Like Whewell, Maury was a polymath, carrying out studies in astronomy, meteorology and, of course, oceanography. In all these fields Maury consistently had a creative mind-set about information and how to get it. Men of science like Whewell were emerging as lofty scholars who made meticulous observations and experiments. For Maury, if a mystery could be solved by dusting off old records, or collected by adding to sailors' orders, then he'd make it so. And he encountered mystery after mystery that benefited from these approaches, not just the one-time "great tide experiment" that struck Whewell.

For example, in 1846, just two years after the Naval Observatory opened, both a French and a British astronomer independently discovered Neptune. Neither actually observed it, but both figured out that the only plausible explanation as to how Uranus's orbit could mathematically disobey the laws of Isaac Newton was the presence of a nearby large planet tugging on it. While the Europeans bickered over credit and naming rights, none thought to compute Neptune's orbit. The best estimates were that it would take half a century to get enough glimpses of it to make the calculation of its orbit possible. Maury saw that there was no need to wait fifty years into the future when one could travel fifty years into the past. He assigned one of his assistants, Sears Cook Walker, to help make observations from the Naval Observatory in order to coarsely approximate the orbit and then trace it backward through historical records snooping for clues. In 1847 they uncovered that Neptune had indeed been seen in 1795 but misidentified as a star (a mistake Galileo had also made in 1613). After a few precisely timed observations, Maury and Walker were able to compute Neptune's orbit, which takes an extravagant 165 earth years to complete its circuit around the sun.

A few years before the Neptune observation, as a navy lieutenant in charge of the Depot of Charts and Instruments, Maury spotted the value of the growing stockpile of old ship logs. What better way to make navigation decisions about a journey than from all previous journeys? With hired helpers he turned the information logged from each vessel's voyage into data on wind and currents. Compiling the glut of data took five years and finally, in 1847, he released the first few sheets of the Wind and Current Chart of the North Atlantic.[111] Maury justified the chart in a letter to former president John Quincy Adams by explaining the need "to generalize the experience of navigators in such a manner that each may have before him, at a glance, the experience of all."

Maury's first efforts were called track charts, and they began a series of specialized charts: pilot charts (starting in 1849), thermal charts (1850), trade wind charts (1851), whale charts (1852), and storm and rain charts (1853). Maury announced the whale charts in spring of 1851, six months before Herman Melville published *The Whale* (later retitled *Moby-Dick; or, the Whale*), which has a footnote about Maury's charts.

As he sent these charts of trade winds, ocean temperatures, and prevailing currents to mariners, he also sent special logbooks and instructions so they could share relevant observations from their current journeys in standardized formats. Because sailors experienced improved navigation and safety when using Maury's charts, more and more signed up for the program, and all benefited from continued refinements and improvement of the charts. By 1851, more than a thousand ships, both merchant and naval, in every blue ocean, were sending reports to Maury.[112] To expand further, he arranged an international maritime conference in Brussels in 1853. Soon after, he began receiving data

---

[111] Terrestrial areas are represented with maps, while oceanic areas are represented with charts.

[112] Maury didn't stop once he hit shore. In the late 1840s, riding on the successes of his mariner data collection program, Maury pushed for an analogous system on land—with farmers. Like Thomas Jefferson, he wanted to start a crowdsourcing system of meteorological observations on land, using the telegraph to aggregate reports in one location where his office could formulate weather forecasts. The Revolutionary War interrupted Jefferson's plans; for Maury it was the Civil War.

from thirteen nations, with ships acting like moving weather stations that chronicled meteorological observations while at sea. Maury reflected, "Though they may be enemies in all else, here they are friends. Every ship that navigates the high seas with these charts and blank abstract-logs on board may henceforth be regarded as a floating observatory—a temple of science."

Before Maury's charts, ocean travel from New York to San Francisco took over a year. With the help of Maury's charts, the trip was possible in a swifter vessel and reduced to a speedy three months. His routes saved lives as well as millions of dollars in ocean commerce. One mariner wrote to Maury, "Until I took up your work, I had been traversing the ocean blindfolded."

Whewell's and Maury's efforts at citizen science are still relevant today. Though Whewell's project lasted only two weeks, global climate change has sparked a revival of projects to document tide marks, in this case to track rising sea levels; this reincarnation of Whewell's "great tide experiment" is community driven. Several organizations in coastal regions of the United States rally residents to document king tides through photographs. Right now king tides (above-average high tides) are the highest tides of the year, but with global climate change, they could become average tides; as such, observations that communities gather about king tides inform policy makers and urban planners in adapting cities to rising sea levels. Ideal photos include some sort of structure, such as a jetty, pier, or dock, which gives an indication of the water line. Participants upload their photos to social media with the date, time, geotag, and hashtag #KingTides. If Whewell's project had not lasted only a mere two weeks but had involved ongoing monitoring, imagine what mysteries of sea level rise would already be revealed by now!

Just as Maury pieced together information by digging through tattered astronomy notes and frayed logbooks, today's citizen science invests in these same tasks. Global climate change makes historical records even more important, maybe even more valuable than they were in Maury's day. Yet accessing historic data is even more challenging today because the logbook stockpile is enormous. In the United

Kingdom alone, there are an estimated 250,000 logbooks with information that could help climate scientists. From billions of observations from ships crisscrossing the seas, researchers are reconstructing historic weather and ice floes.

The Smithsonian Transcription Center is an online hub of projects where people can transcribe historical ledgers, logbooks, manuscripts, photo albums, and specimen labels. Although scanning machines with optical recognition software can make digital copies of old paper documents that were typed, only a human eye can consistently decipher the idiosyncratic features of handwriting. The digital deckhands at the Smithsonian call themselves "volunpeers" to emphasize that though they are unpaid volunteers, their interest and intellect are on equal footing with those of professionals. Meghan Ferriter, the coordinator of the Transcription Center, coined the term *volunpeer*, which allows scientists and citizen scientists alike to thumb their noses at Whewell's concept of "subordinate labourers." In the transcription project Old Weather, which sprouted from Zooniverse (see chapter 4), volunpeers transcribed over 500,000 handwritten pages of logbooks from commercial whaling ships that ventured into Arctic waters.

When volunteers transcribe old logbooks they recognize, as Maury did, that what appears to be minutiae on each page is, collectively, scientifically powerful. Scientists, like Kevin Wood at the Joint Institute for the Study of the Atmosphere and the Ocean at the University of Washington, are reconstructing sea ice conditions of the past. Understanding these weather and climate variations will help Wood and his colleagues predict what's in store for our future.

Maury's citizen science project, in which sailors shared observations to improve navigation, never stopped (except briefly during the Civil War); it continues to this day in the form of *Sailing Directions*, a forty-two-volume publication from the National Geospatial-Intelligence Agency (NGA). *Sailing Directions* is based on data contributed by merchant ships and has been online since 2005. The NGA is credited with gathering the intelligence that allowed the US military to raid Osama bin Laden's hiding place in Pakistan in 2011. No one credits the NGA with citizen science; the NGA calls Maury's brainchild tradecraft be-cause the observations, though not sent as messages in bottles or passed via dead drops, function like geospatial intelligence to help the United States on issues of national security. The NGA motto "Know the Earth—Show the Way" characterizes Maury's philosophy. The daily routine of recording standardized data while sailing is embedded in the customs and responsibilities of ocean travel.

I call Maury's legacy citizen science because his methods were proof positive that scientists and people with other jobs could cooperate to-ward the goal of making knowledge that is beneficial to us all. We sink or swim together.

The stories of Maury and Whewell are like folklore to me now. Whewell's masterminding of "subordinate labourers" reminds me of Tom Sawyer duping others into whitewashing a fence—but in this case along the changing tide marks.[113] To be fair, participants were not hood-winked, like Tom Sawyer's friends were, and certainly people wanted to help an esteemed scholar such as Whewell. But seeing how citizen science holds the capacity to benefit more than the scientist's agenda, maybe people in those maritime communities were in fact cheated just a little.

Maury's strategy, on the other hand, embodies the legend of "Stone Soup," the old folktale that teaches the lessons of collaboration for im-proving the conditions of everyone in a village. In the story, a trio of monks traveling through the war-torn countryside arrive in a quiet town. They set up shop in the center plaza, which is devoid of hustle and bustle because distrustful villagers stay close to home. The monks drop several stones in a large pot of water and set it to boil, then an-nounce that stone soup will be tonight's dinner for everyone. Although the villagers are skeptical at first, curiosity and the desire for a better dinner soon win them over. No one villager possesses enough on his or her own for a complete meal. But, together, they do. In turn, each de-

---

[113] An even more devilish tale would be to see Whewell as an uncharitable Little Red Hen: "You helped plant the wheat, harvest the wheat, thrash the wheat, and make the flour, but I'm going to eat the bread."

cides to visit the monks and share a little: "I can spare a few carrots for the soup"; "Here are some potatoes"; "I can share some spices." Through cooperation and sharing, the entire village feasts on delicious, nutritious soup.

Citizen science utilizes the whole cupboard of Stone Soup recipes. Scientists unfurl a large blank database with the promise of discoveries; researchers may fill in a few observations, and then hobbyists from across the world each share what they have seen. Not everyone contributes equal amounts, but that's all right. A relative few may share an abundance of the base stock of potatoes, carrots, and leeks; many others may barely toss in a shake of salt or pepper; separately these observations do not necessarily provide big insights, but together—when we stir and simmer the contributions, big and small—we can develop a shared understanding of the world. Maury created a system through which sailors could chip in their observations, and together they revealed patterns of ocean winds and currents. Today information and communication technologies bring this soup to a boil more quickly and add sophistication and legitimacy. The volunpeers of today are transcribing the logbooks of yesterday because observations become even more valuable with time. Not long ago, the word *hobby* was almost a pejorative, and *show-and-tell* was confined to elementary school. These were overindulgences that we were supposed to outgrow. Now they are purposeful means of discovery.

Citizen science is not a hammer where every problem is a nail. But it has no equal for scientific research that requires long timescales, coverage over large geographic and oceanographic areas, access to residential systems, and lots of—or strategically placed—eyes on the lookout for rare phenomenon. It brings minds together to outsmart computers. It is vital to decision making for personal health and the management of natural resources. It is indispensable in balancing the scales of environmental justice. From studying the last decade of citizen science, we now know that it cocreates highly reliable scientific knowledge *and* weaves social capital. Conventional science is a half measure; it is citizen science that holds both of the two keys for problem solving.

From the roots of social capital sprout solidarity and reciprocity, encapsulated by the ideal "I'll scratch your back if you scratch mine." Trust and cooperation branch from it. "Art is I; Science is We" wrote Claude Bernard,[114] a French physiologist in the 1800s. We can each gaze at a painting and experience it uniquely; art allows different meanings from individual perspectives —*I*. Yet when we gaze at the sunset, we should each come to agree that the sun appears to set because the earth is rotating. Science provides a shared understanding of what's observed—*we*. Citizen science is still in its infancy and, if we grow it, we have an opportunity to develop systems of engagement and participation aimed at collective problem solving. Citizen science is *us*.

The rise of the Internet ignited concern that our society was losing social capital. As Robert Putnam explored in his book, *Bowling Alone: The Collapse and Revival of American Community*, the 40 percent drop in organized bowling leagues between 1980 and 1993 was illustrative of a larger trend. Voting in congressional elections is even less popular than bowling. Americans invite friends over less often. Families often don't eat dinner together. Many long-standing social traditions are falling by the wayside.

Maybe social capital is not being lost but simply changing its appearance. We may bowl alone, but together we count trash, measure air quality inundated with pig farts, and manage our health. The growth in citizen science may be a sign of a growth in social capital.

Social scientists have found that communities rich in social capital achieve higher educational levels and have better governments, stronger economic growth, and less crime. People in communities with social capital are happier and healthier. When we are connected, we are more likely to have each other's backs. We can tackle problems together.

Sailors received charts that helped their navigation. Birders download maps that help them find birds. When everyone gives a little, we all gain a lot. Aside from learning, people add meaning to their lives, make last-

---

[114] Bernard engaged in vivisection (conducting experiments on living animals), which motivated his wife (whose dowry funded his research) to obtain an official separation in 1869 and then actively campaign against the practice.

ing friendships, and create space in which to follow their passions. The maxim that people protect what they care about seems to hold up. Citizen science heightens our awareness and how much we care about what we observe. That's why we have dark-sky parks, bird-friendly coffee, and people protecting roadsides as butterfly gardens. Collectively understanding the world leads people to cooperate to improve the world.

The stories in this book are a small sampling of citizen science, but you can find thousands of citizen science projects on SciStarter.com. There were many fields and examples that I could not cover in this book, notably projects that use volunteer geographic information for governance, planning, and disaster management, and movements that intersect with citizen science—those of makers, smart cities, and the quantified self. There are citizen scientists who are innovators and explorers, like David Lang and Eric Stackpole, who designed an undersea drone because they wanted to find lost gold in an underwater cave. Now they sell low-cost, robotic, undersea vehicles (with open-source software, hardware, and electronics) that give people the ability to unlock the mysterious fathoms of the ocean for exploration. There are groups of people—particularly those far from where I live—that I could not cover in this book: environmental justice projects in India, giraffe monitoring by the Massi in Tanzania, volcano monitoring in Peru, and marine surveys in Israel to name a few.

Wikipedia launched in 2001, and by 2015 it contained over thirty-five million articles in 288 languages written by fifty-five million registered users and many anonymous contributors. Wikipedia has prepared society to expect almost instantaneous access to valid information sourced from ordinary people. Wikipedia and social media have moved us from reliance on paid experts to accepting the value and wisdom of crowds. The future for citizen science is in a society of people holding an even greater expectation: access to *systems in which ordinary people advance valid new discoveries*.

For too long, society has had a love-hate relationship with science.

We love its technological offspring: refrigerators, light switches, thermometers, and the darling miniature computers we hold in the palm of our hands. We love the medical breakthroughs: the stents, joint replacements, and cough drops. At the same time, we hate the seeming recklessness of science. Propelled beyond caution, either by arrogance or unbridled curiosity, inventions from the latest science unearth moral dilemmas: cloning, drones, the atomic bomb.

The society-versus-science conflict is like a larger reflection of a conflict in human nature between our inborn desire for safety and our innate curiosity. Our strong desires for both safety and curiosity may explain why surveys show that the professions that yield the most respect for bettering our world are those of soldier, teacher, and scientist. Our yearning for safety urges us to stick with what we know, to stay in our comfort zone, to hold to the status quo. Yet our curiosity propels us toward novel experiences, learning, and change.

When science and society collide, injustice ensues. We split into intellectual haves and have-nots. The haves feel ownership of science and attempt to wield its power to their advantage. Like one end of a societal seesaw, scientists can exaggerate the importance of data by placing it at the center of decision-making. The have-nots feel alienated from science and attempt to deny its power; on the have-not side of the seesaw, the public tends to undervalue data, dismissing complexity and inconvenient information. I have come to see hobbyists, amateurs, and civically engaged people who participate in scientific research as gathering at the center, demonstrating balance on the beam that is the fulcrum of the seesaw. For this reason alone, citizen science has power and potential greater than that of science confined to the ivory tower.

I chose to end with Maury because he was a public scientist, because his legacy continues and, most of all, because we are all like sailors, balancing adventurous curiosity with the desire to safely make it home. Maury's citizen scientists were sailors in navies and merchant mariners; they were ordinary people. They observed to make their journeys safer and shared those observations to make the journeys safer for those following in their wake. Observing and sharing became intertwined with the responsibilities of being a sailor, sharing the high seas with mariners to come.

How do we look after the safety of future generations who will make their journeys around the sun after our ships have sailed? Observing and sharing our observations will become what it means to be a responsible human residing on planet earth. Whether you are a birder, a weather bug, a stargazer, a beachcomber, a patient, an activist, a thinker, a tinkerer, or a quantified selfer, this type of tradecraft is about noticing and sharing—without secrecy. Our planet is a ship orbiting the sun, orbiting the center of our galaxy, riding each wave into the future. There is so much that we do not understand about our world, yet every day we must make decisions to navigate our path forward. Citizen science is a passport to the rights and responsibilities of engaging in validated systems of discovery.

With citizen science, we no longer sail blindfolded.

INTERVIEWS/SOURCES

INTERVIEWS

Allee, Leslie, Ithaca, NY, November 2014

Baker, David, via phone, March 2015

Brody, Julia, via phone, January 2016

Coleman, Bill, via e-mail, October 2014

Crick, H. Q. P., via e-mail, September 2012

Danielsen, Finn, via e-mail, June 2015

Doesken, Nolan, via phone, October 2014

Dunn, Raleigh, NC, July 2012

Edwards, Marc, Raleigh, NC, April 2015

Edwards, Marc, via phone, September 2016

Fahey, Nancy, Wrightsville Beach, NC, July 2012

Fortson, Lucy, St. Paul, MN, September 2015

Fortson, Lucy, via Skype, September 2015

Gould, Andrew, via e-mail and phone, May 11, 2015

Grant, Gary, via e-mail, January 2013

Grant, Gary, Chapel Hill, NC and Tillery, NC, January 2013

Herring, David, Richmond, VA, June 28, 2012

Hiiragan (pseudonym), via e-mail, 2013

James, David, via phone, October 23, 2014

James-Creedon, Jackie, via e-mail and phone, December 2015
and January 2016

Kientz, Vivian, via e-mail and phone, October 2014

Lamerichs, Sebastian, via e-mail, 2013

Ludwig, Ted, via phone, February 2016

Norris, Shelia, via e-mail, November 2014

Menninger, Holly, Raleigh, NC, July 3, 2012

McCormick, Jennie, via e-mail and Skype, October 2015

Miller, Susan, Wrightsville Beach, NC, July 2012

Haklay, Mordecai ("Muki"), London, UK, 2012

Haklay, Mordecai ("Muki"), via e-mail, 2012

Muhammed, Naeema, Chapel Hill, NC and Tillery, NC, January 2013

Ries, Leslie, Ithaca, NY, Spring 2014

Roy, Helen, via e-mail, October 2014

Rolfes, Anne, via e-mail and phone, July 2015

Robinson, Matthew, via phone, June 2016

Seaton, Keith, via e-mail, August 2014

Shamblin, Brian, via e-mail, October 2014

Swendel, Ann and Scott, via e-mail, July 2015

Sullivan, Brian, via e-mail, 2016

Takada, Hideshige, via e-mail, Sept 2012, 2014

Taylor, Ginger, Wilmington, NC, July, 2012

Thomas, Bill, via e-mail, August 2014

Walters, LeeAnn, via phone, September 2016

Warden, David, via e-mail, September 2014

Wesley, Doug, via phone, October 2014

Wangchuk, Tshewang, Raleigh, NC, February 2015

Wilbanks, John, via phone, January 2016

Wing, Steve, Chapel Hill, NC, January 2013

SOURCES

Amagoalik, Simeonie. Nunavut Climate Change Centre. http://climate changenunavut.ca/

Liebenberg, Louis. *The Art of Tracking: The Origin of Science*. Claremont, South Africa: David Philip Publishers, 1990.

Dan Maclean et al. "Crowdsourcing Genomic Analyses of Ash and Ash Dieback – Power to the People." *GigaScience* 2, no. 3 (February 12, 2013).

Mike, Elaiya. Nunavut Climate Change Centre. http://climatechange-nunavut.ca/

Peart, Roger. "Volunteer Stories." British Trust for Ornithology. https://www.bto.org/volunteer-surveys/taking-part/volunteering/volunteer-stories.

Swartz, Aaron. "Who Writes Wikipedia?" Raw Thoughts. http://www.aaronsw.com/weblog/whowriteswikipedia.

WarerontheMoon (pseudonym), Straub, Miranda C.P. "Giving Citizen Scientists a Chance: A Study of Volunteer-led Scientific Discovery." *Citizen Science: Theory and Practice* 1, no. 1 (2016): 5.

CALL TO ACTION

IF YOU WERE INSPIRED BY THE STORIES IN THIS BOOK, THEN I HOPE YOU are wondering: "What's next?"

There are many ways to get involved in citizen science, or more deeply involved than you already are. Citizen science is rapidly growing. That's great news—but it can also be confusing and overwhelming. It isn't easy to navigate the maze of old and new projects, many of which concern similar topics. Adding to the complexity is that some projects and resources are unexpectedly temporary, and some are specifically local or regional. Where are you needed? Where will you be welcomed? Where can you lend a hand to what's of most concern to you?

This book is static, but projects of citizen science are dynamic—it is an ever-changing landscape. Rather than compile a list of projects and resources that would soon be out-of-date, I want to point you to a one-stop-shop, the Amazon of citizen science choices. It's called SciStarter.

For the past several years I've been working as part of an amazing crew at SciStarter.com, led by Darlene Cavalier (the founder of the Science Cheerleaders who you met in chapter 6). With help from the National Science Foundation we've been building new features and functionality to this website. As this book goes to press, we have over 1,500 citizen science projects in the SciStarter ecosystem—that makes SciStarter the largest collection of citizen science projects in the world. Project owners add their projects to SciStarter for free; citizen scientists join for free. We help them find one another. As you begin to explore the world of citizen further, you can easily keep track of the projects you've joined, contributed to, or simply bookmarked through your personal SciStarter dashboard.

Enter and join via website: **www.SciStarter.com/Cooper** and you'll be tagged as my guest.

I want you to start your voyage into SciStarter from my landing page for two reasons: first, so that I can welcome you to the site and guide you in how to use this resource to navigate the evolving world of citizen science; and second, I want to know you as a budding citizen scientists, in a research sense. Go to my landing page as your path to joining SciStarter and YOU can be part of a big experiment, part of the cohort that came to SciStarter after reading this book. We want to see what you do, what projects you like, and learn what we can from you in order to help you have the best citizen science experiences imaginable.

A SPECIAL CALL TO TEACHERS AND EDUCATORS

Teachers help students grow into well-rounded adults. In school, students learn art and sports, which become career paths for a few and meaningful hobbies for most. Similarly, even without choosing science as a career path, students can learn how to participate in making new discoveries through citizen science. When teachers can offer authentic, hands-on science experiences, students can gain a love for discovery and a lifelong curiosity to understand not only what is already known but also what is not yet know. As a form of civic engagement and volunteerism, citizen science is useful in classes other than science too.

A one-stop-shop of resources for bringing citizen science into schools and classrooms is Students Discover: **www.StudentsDiscover.org**

At Students Discover, the lesson plans are freely shared for K-12 and increasingly for university classrooms too. Each lesson plan was developed collaboratively among teachers, scientists, and design teams and align with Next Generation Science Standards and/or Common Core Standards.

ACKNOWLEDGMENTS

TO ALL THOSE WHO HELPED ME BEGIN, STICK WITH, AND FINISH THIS BOOK, I hope you find pleasure in reading it.

To my husband, Greg Sloan, for patience with my work-life imbalance, encouragement through my self-doubts, and for not mansplaining Einstein for chapter 4.

To my dad, Jay Cooper, because this book would not have been possible without his establishing the expectation that I could do it, plus his edits and encouragement every step of the way.

To my mom, Barbara Cooper, for teaching me the art of spotting the gems in thrift stores. Turns out that the skill is transferable to science, writing, and more. Now I can spot the gems in life wherever I am.

To the Environmental Leadership Program for guiding me onto the path of writing this book, and Karen Purcell who listened patiently and then impatiently prodded me into action.

To former colleagues at the Cornell Lab of Ornithology, including Colleen McLinn for modeling how to pursue one's career fearlessly, and Rick Bonney and Tina Phillips for showing me how fascinating the people were who provided the numbers about birds that I was busily crunching for my research.

To Dan Decker, for mentoring me in the social sciences and believing this book was a good idea.

To Steve Strogatz for seeing the potential for a book about citizen science before I saw it and in assuring me that it is respectable for scientists, irrespective of their job title, to write books.

To Hugh Powell for reading very early and poor drafts and kindly not destroying the blossoming idea. To Marc Messing for helpful critiques of later drafts. To Karina Knoll for advice about finding a publisher, Bruce Lewenstein for telling me about Matthew Fontaine Maury, as well as lots of advice and insights, and Paul Gray for telling me about Whewell's Great Tide Experiment. To Dave Leech for putting me in touch with citizen scientists with the British Trust for Ornithology.

To Soledad Exantus and Anne Jacobson for our east coast road trip in Soledad's van with all of our daughters. That trip included interviewing David Herring (chapter 1), the sea turtle monitors at Wrightsville Beach (chapter 7), and my first blog post at *Scientific American*.

To my friends Mark and Deirdre Silverman for encouragement, reading drafts, and letting me use their home for writing retreats.

Special thanks to Darlene Cavalier (chapter 6) for my first steady blogging opportunity and making the world of citizen science in general, and my corner of it, shine bright every single day.

To Geoffrey Haines-Stiles for sharing photos from *The Crowd & The Cloud* (CrowdAndCloud.org) and providing edits and feedback.

To Holly Menninger and Rob Dunn who I first met when I interviewed them for this book (chapter 6). I am grateful to them for designing excellent citizen science and garnering attention to it until NCSU and the NC Museum of Natural Science decided to invest in this frontier.

To Bora Zivkovic for introducing me to science blogging, and his steadfast encouragement about the importance of science writing.

To Billy Tusker Haworth for many discussions that helped sustain my curiosity and excitement about citizen science.

To Sandra Steingarber for cheering me on and serving as my inspiration for finding Charlotte Sheedy, who directed me to my ever-helpful and level-headed agent, Mackenzie Brady Watson. Thanks, Mackenzie, for having my back throughout this process. I thank my first editor, Dan Crissman, for believing in this project and suggesting to organize this book by disciplines. I would not have gotten this book off the ground without his vision. I thank Vanessa Kehren and Allyson Rudolph for helpful edits and discussion, and Chelsea Cutchens, for shepherding this book through to completion.

And lastly, to all the scientists who helped me understand their discipline better and all the citizen scientists who helped me understand their experience better. For those who helped but do not appear in the book, I want to offer thanks to Wouter Buytaert (water monitoring in Peru), Rinjan Shrestha (snow leopard monitoring in Nepal), Renaldo Browne (Firefly watcher), Eleanor Starkey (River Watch in UK) and Pat Foreman (RiverWatcher in UK), Michael Heimbinder (HabitatMap and Aircasting), Derya Akkaynak (Divers4Oceanography), David Wald (Did You Feel It?), and Aitana Oltra (Mosquito Alert). I wish I could have included everyone's story.

INDEX